SURVEY OF INSTRUMENTS FOR MICROMETEOROLOGY

IBP HANDBOOK No. 22

Survey of Instruments for Micrometeorology

COMPILED BY

J. L. MONTEITH

Department of Physiology
and Environmental Studies
University of Nottingham

INTERNATIONAL BIOLOGICAL PROGRAMME

7 Marylebone Road London NW1

BLACKWELL SCIENTIFIC PUBLICATIONS

Oxford London Edinburgh Melbourne

and Published for them by
Blackwell Scientific Publications
Osney Mead, Oxford,
3 Nottingham Street, London W1,
9 Forrest Road, Edinburgh,
P.O.Box 9, North Balwyn, Victoria, Australia.

ISBN 0 632 08520 7

FIRST PUBLISHED 1972

Distributed in the U.S.A. by
F.A.DAVIS COMPANY, 1915 ARCH STREET,
PHILADELPHIA, PENNSYLVANIA.

Printed in Great Britain by
BURGESS & SON LTD.,
ABINGDON, BERKSHIRE.
and bound by
THE KEMP HALL BINDERY,
OXFORD.

Contents

Foreword

The International Biological Programme is a world study, lasting from 1964 to 1974, of 'biological productivity and human adaptability'. It is divided into seven sections which deal with terrestrial productivity, production processes such as photosynthesis and the nitrogen cycle, terrestrial conservation, freshwater productivity, marine productivity, human adaptability and the use and management of biological resources. Previous handbooks in this series have mainly been concerned with the methodology of a specific section. This one, however, is intended for use by the whole programme because the techniques are necessary for all. They may be less important for studying aquatic ecosystems but even with them the air-water interface and the atmosphere immediately above it has high significance and many aquatic organisms spend part of their life cycle in the air or on the adjacent land.

This handbook is different from the rest in another respect in that the source of the information which it contains is mainly from the manufacturers of instruments. It is not long since the biologist who needed to measure micrometerrological variables had to make his own instruments but this list shows that today there are many on the market.

The greater part of the task has been undertaken by John Monteith, formerly of Rothamsted Experimental Station and now Professor of Environmental Physics at the University of Nottingham. He is one of two officers appointed by the Special Committee for the IBP for liaison with the WMO, and IBP's thanks are extended to WMO for ensuring that, in preparing this handbook, such liaison has been close.

E. B. Worthington

Introduction

Micrometeorology is the study of physical processes in the atmospheric boundary layer, usually within a few metres of the ground. Meteorologists are concerned with the behaviour of the boundary layer because it determines the input of heat, water vapour and momentum to the lower atmosphere and a large number of instruments have been developed for micrometeorological measurements. Many of these instruments can be used by ecologists to measure and define the environment of plants and animals and to explore the ways in which organisms modify the environment they are exposed to. In English-speaking countries, the measurement and description of biological environments is usually referred to as 'microclimatology' but the term acquires different shades of meaning when it is translated into other languages. The more general term 'micrometeorology' was therefore chosen for the title of this handbook.

Twenty years ago, ecologists seeking advice about micrometeorological equipment were usually advised to design and make their own. Manufacturers have now recognized that there is a large market for instruments that can be used either in physical studies of the lower atmosphere or in ecological projects. A wide range of instruments is available commercially and many of these have been developed from research prototypes. The evolution of production models from prototypes has often run into trouble because manufacturers have tended to forget that a micrometeorological instrument must be immune from the depredations of animals and birds as well as from the effects of wind, rain and prolonged sunshine. Standards are rising, however, and most instruments satisfy their formal specifications. So many new instruments have appeared on the market in the last 5 to 10 years that it is becoming increasingly difficult to obtain and compare descriptions of all the similar types of sensor or recorder that are available. As recently as 1960 for example, the choice of instruments for measuring solar radiation was limited to several rather cumbersome models used by meteorological services. The range is much wider now and nearly 30 instruments of this type are described in the handbook.

Several sections of the International Biological Programme have been specially concerned with meteorological and micrometeorological measurements. The Production Processes section (PP) includes projects on the distribution of radiation in vegetation and its relation to photosynthesis rate. Other sections are concerned with the energy and water balance of ecosystems and with the relationship between microclimate and human comfort. Brief references to meteorological measurements can be found in some of the handbooks already published in this series (see p. 260) reviewing techniques for a specific aspect of biological research. This handbook meets a different type of need: it is primarily intended to guide the choice of instrumentation. Advice on how to make, calibrate, and use micrometeorological instruments can be obtained from the papers and books listed in the references and bibliography which cover the period 1960-1970. A few of the instruments described in the literature are available commercially, often in a somewhat modified form, and cross-references have been included in the manufacturers' specifications.

The committee responsible for planning the handbook began by devising a standard form on which manufacturers could describe the main features of a sensor or recording system. Correspondents in four regions – Europe, North America, Australasia and Japan – were then asked to distribute the forms to manufacturers whose instruments were known to have been used in micrometeorological studies. Many firms returned these forms completed to the editor and their cooperation is recognized by an asterisk beside the name of the company. Other firms sent their standard literature from which it was usually possible to construct a skeleton specification,but most glossy leaflets seem to be designed to attract rather than to inform potential customers. Some firms which produce micrometeorological instruments may not be represented in the handbook either because they failed to return a questionnaire or because correspondents had no information about them. Omissions of this kind are regretted. The editor will be pleased to hear from manufacturers producing relevant equipment which has not been included in the handbook as well as from authors who have published papers that were overlooked when the references were compiled. It is hoped to publish this additional material either in the form of a supplement to the first edition or in a second edition.

A wide variety of units was used by different manufacturers and these have been retained (except in a few specifications which contained metric and English units or °C and °F side by side!) A table of approximate conversion factors is on page x. Many completed questionnaires were rejected because they described equipment that was outside the scope of the handbook; in particular, standard climatological instruments used by meteorological services which are usually not appropriate for micrometeorological work.

Although all the entries in the handbook have been carefully checked, the editor and publishers accept no responsibility for mistakes in the description of any instrument. It would be unwise to order equipment from a manufacturer solely on the basis of the abbreviated description published here. Apart from the danger of inaccuracies, the design of micrometeorological instruments is frequently improved in the light of field experience. Manufacturers should therefore be consulted before an instrument is ordered.

I am glad to acknowledge help from the large number of people who contributed to the production of the handbook: from the committee which conceived and planned it – Dr. K.L. Blaxter, Dr. J. Bligh, Dr. J.P. Cooper, Dr. O.G. Edholm, Dr. N.E. Rider and Mr. L.P. Smith; from the correspondents who prepared lists of manufacturers and circulated questionnaires; from manufacturers who completed the questionnaires or sent collections of literature. I am also grateful to Lionel Smith for supplying many of the references from the bibliography which he edits for 'Agricultural Meteorology'; to Mrs. Susan Kirby of the Royal Society who was responsible for all the initial correspondence; to Mrs. Dirkje Trivett who collated references and typed the manuscript; and to my son David for help with arranging the material for publication and with correcting proofs. The front cover, illustrating one of the problems of micrometeorological measurement, was designed by Bob Gibson.

May 1971

J.L. Monteith,
School of Agriculture,
University of Nottingham,
Sutton Bonington,
Loughborough,
Leics., U.K.

Abbreviations and approximate conversion factors

Length

m	metre		= 3.3 ft
cm	centimetre	= 10^{-2} m	= 0.4 in
mm	millimetre	= 10^{-3} m	= 0.04 in
μm	micrometre	= 10^{-6} m	
nm	nanometre	= 10^{-9} m	
ft	foot		= 0.3 m
in	inch		= 2.5 cm

Mass

kg	kilogramme		= 2.2 lb
g	gramme	= 10^{-3} kg	= 0.035 oz
lb	pound		= 0.45 kg
oz	ounce		= 28 g

Time

h	hour
min	minute
sec	second

Temperature

°C	degree Centigrade	$=5(F-32)/9$
°F	degree Fahrenheit	$=(9C/5)+32$

Power per unit area

W/m^2	watt/square metre	$= 0.1\ m\ W/cm^2 = 0.0014\ cal\ cm^{-2} min^{-1}$
$cal\ cm^{-2} min^{-1}$	calorie per square centimetre per minute	$= 700\ W/m^2$

Electrical units

V	volt	
mV	millivolt	$= 10^{-3}$ V
A	amp	
mA	milliamp	$= 10^{-3}$ A
μA	microamp	$= 10^{-6}$ A
Ah	ampere-hour	
k ohms	kilohms	$= 10^3$ ohms

Cost

£	pound sterling	$2.55 (Jan. 1972)
$	dollar American	£0.39 (Jan. 1972)
$A	dollar Australian	£0.47 (Jan. 1972)

Other abbreviations

vpm	volumes per million, e.g. m^3 of CO_2 in 10^6 m^3 of air
fs	full scale
rh	relative humidity
DC	direct current
AC	alternating current
*	specification provided directly by manufacturer; other specifications were abstracted from manufacturer's literature.

Keys

Each group of instruments is arranged in alphabetical order of manufacturers.

Sensors (pages 1–201)

1 Type of instrument and manufacturer's identification

2 Manufacturer's name and address

3 Specification
a principle of operation
b size
c weight
d form of output
e range or sensitivity
f response time (i.e. time for 63% of signal change unless otherwise stated)
g accuracy and resolution
h working range
i length of cable
j approximate cost (1971)

4 Ancillary equipment or power supply

5 Presentation of output
a meter
b recorder

6 Other features

7 Users

Recorders and Integrators (pages 203–236)

1 Type of instrument and manufacturer's identification

2 Manufacturer's name and address

3 Specification
a principle of operation
b size
c weight
d range or sensitivity
e accuracy and resolution
f stability
g chart width and speeds
h number of channels and cycle time
i power needs
j working range
k approximate cost (1971)

4 Other features

INSTRUMENTS

1
Thermometers and psychrometers

A **thermometer** is any type of sensor which is designed to measure the temperature of its surroundings. A **psychrometer** is a pair of thermometers one of which is covered with a wet sleeve. Atmospheric humidity is uniquely related to the wet and dry bulb temperatures of a properly ventilated psychrometer.

Mercury in glass thermometers are seldom used in micrometeorological work because they are relatively bulky and cannot be read from a distance but the Assmann psychrometer is often used a standard against which other thermometric or psychrometric systems can be checked. **Thermocouples** are very cheap to make and are available commercially but have the disadvantage of needing a reference junction and giving a relatively small output, e.g. 40 μV/deg C for copper and constantan. After a period when thermocouples were displaced by other types of thermometer, they are regaining favour in conjunction with stable solid-state D.C. amplifiers. **Resistance thermometers** with nickel or platinum wire are more difficult to make but miniature types can be bought. **Thermistors** are small beads of a semi-conductor with a large negative temperature coefficient which decreases with increasing temperature. Modern thermistors are very stable and can be closely matched for operation in a common bridge circuit. Thermistor probes are available in many shapes and sizes and are extremely useful for spot measurements in plant, animal or human studies. **Junction diodes** have recently been used for temperature measurement (Sargeant and Tanner, 1967): the voltage across a forward biassed diode decreases linearly at about 2.3 mV per deg C. Only one description of a commercial diode thermometer was submitted for the handbook. **Quartz crystal** thermometers are capable of very high precision but they are very expensive and relatively bulky.

In micrometeorological work it is essential to protect temperature sensors from radiation and to promote heat exchange by a rapid flow of air. Most workers prefer to design their own radiation shields, often with forced ventilation, but several commercial shields are available and examples are included.

Kata thermometers are used mainly in medical studies to determine the rate at which the bulb of a relatively bulky thermometer cools when it has been heated, e.g. to 100°F. The rate of cooling, which depends on the windspeed and temperature of the ambient air, is taken as an index of the 'cooling powe of the air. The temperature recorded with a **black globe thermometer** has been used in many medical and animal studies to indicate the effective temperature of the environment, taking special account of radiative exchanges as well as air temperature and windspeed. In conditions where sweating is important, the globe thermometer temperature is sometimes combined with wet- and dry- bulb temperatures to give a Wet Bulb Globe Temperature Index.

Several firms are now offering systems for integrating temperature with respect to time, and examples will be found in this section. Temperature recording systems are also included.

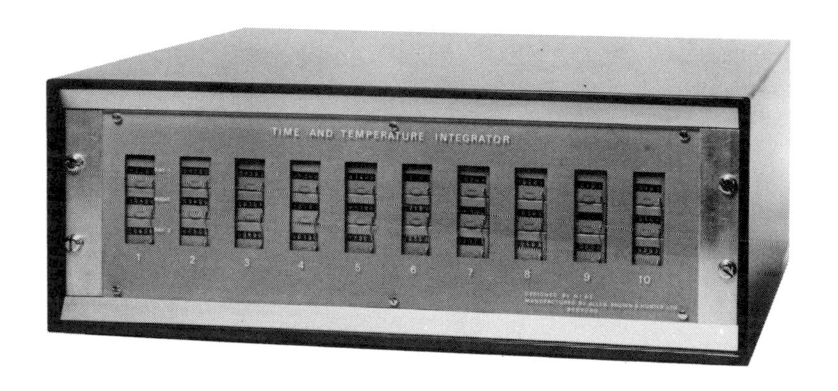

1 Temperature integrator

2 Allen, Brown & Hunter Ltd.
15 Cardington Road,
Bedford, U.K.

3
a Oscillator modulated by thermistor resistance
b 50 x 19 x 31 cm
c 15 kg
d Digital counters
e 5°to 100°C
f Basic cost £53 plus £19 per channel, maximum 10 per unit

4 Power supply 24OV, 50Hz

5
a Set of three digital counters per channel for day-night-day integration switched externally

6 see Weaving, G S. (1970) p. 242

7
a Ministry of Agriculture, Fisheries & Food,
Plant Pathology Laboratory,
Hatching Green, Harpenden, Herts, U.K.
b Ministry of Agriculture, Fisheries & Food,
Lea Valley Experimental Horticulture Station,
Ware Road, Hoddesden, Herts, U.K.
c National Institute of Agricultural Engineering,
Silsoe, Bedfordshire, U.K.

See page xii for Key

1 Kata thermometer

2 C.F. Casella & Co. Ltd.,
Regent House,
Brittania Walk,
London, N.1., U.K.

3
a Cooling of thermometer with large bulb
b 22.5 x 1.9 x 1.0 cm
c 40 g
d scale
e 95°to 100°F, 125°to 130°F, 140°to 150°F
f depends on wind speed and environmental temperature
h unlimited

3 For measuring low air velocities only dry bulb readings are taken but for use as a 'comfort' meter, wet and dry bulb readings are needed. A silk sleeve is provided for wet bulb measurements. (see Bedford, 1964, p. 261)

See page xii for Key

1 Globe thermometer

2 C.F. Casella & Co. Ltd.,
Regent House,
Brittania Walk,
London, W.1., U.K.

3
a Temperature of blackened sphere
b 15 cm diam.
c 0.5 kg
d thermometer scale
e 0° to 65 °C

6 The equilibrium temperature of a black metal sphere is a measure of the effective temperature of the environment, often used in studies of human comfort. (see Bedford, 1964 p. 261)

See page xii for Key

1 Thermocouple

2 Comark Electronics Ltd.,
Littlehampton,
Sussex, U.K.

3
a Thermo-electric effect
b Minimum 2.5 mm diam. x 10 cm long
d Voltage
e -200°to 1100°C depending on type
h As e

5
a A wide range of meters is available reading temperature directly, e.g. from -200°to 30°C in steps of 10°C

6 A range of thermocouples is available

See page xii for Key

1 Resistance thermometer

2 Degussa*
D-6450 Hanau F.R.G.,
Leipziger Strasse 10,
P.O. Box 622, Germany (BRD).

3
a Minimum 2 mm diam. x 25 mm long
d Resistance or bridge voltage
e -60°to 150°C, about 0.4 ohms per deg °C
h -60°to 150°C
j £4

6 A wide range of resistance thermometers is available

See page xii for Key

1 Thermocouple

2 Degussa
D-6450 Hanau F.R.G.,
Leipziger Strasse 10,
P.O. Box 622, Germany (BRD).

3
a Thermo-electric effect
b Minimum 0.2 mm diam. wires, spot welded
d voltage
e -200° to 1600°C depending on type
g Pt – Rh 0.5% tolerance on e.m.f.

6 A wide range of thermocouples is available

See page xii for Key

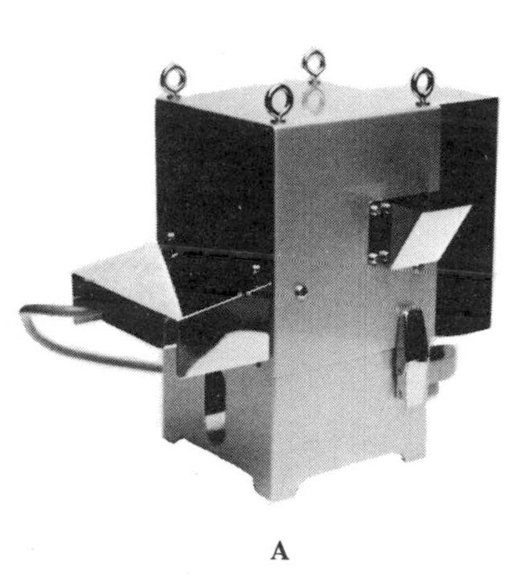

A

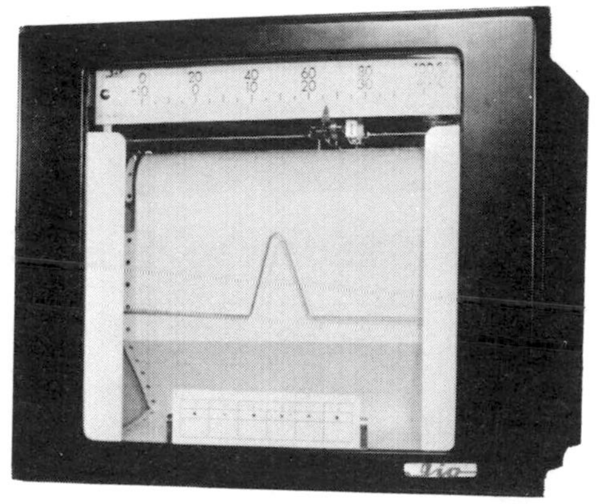

B

1 Aspirated psychrometer with resistance thermometers

2 Iio Denki Ltd.,*
Yoyogi 2-27-18, Shibuya-ku,
Tokyo, Japan.

3
a Change of resistance
b Sensor 20 x 11 x 11 cm (A)
Recorder 30 x 36 x 30 cm (B)
c Sensor 2.2 kg, recorder 20 kg
d Bridge voltage
e Temp. -10°to 40°C, humidity 0 to 100% rh
g Temperature 0.5%, humidity 2% rh
h -10°to 40°C
0 to 100% rh
i 40 m
j $1,280 to 1,670

4 100 V, 50 or 60 Hz.

5
b Incorporated

See page xii for Key

A

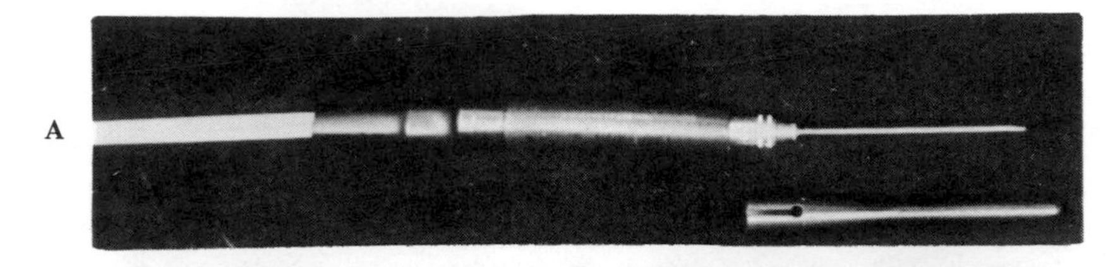

B

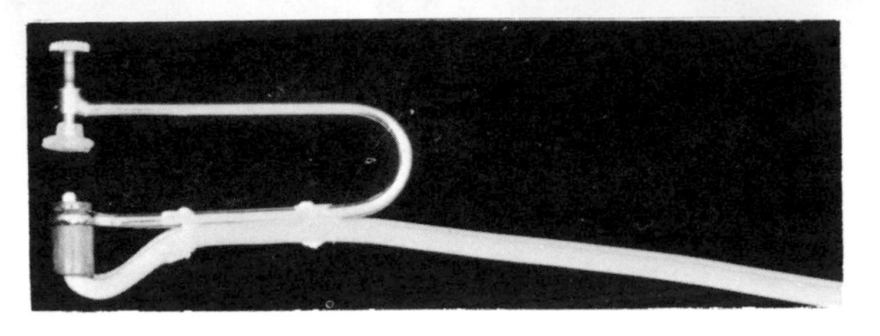

1 Thermistor

2 Iio Denki Ltd.,
Yoyogi 2-27-18, Shibuya-ku,
Tokyo, Japan.

3

a change of resistance

b 0.08 x 3.5 cm (for stem temp. A)
0.2 x 0.2 cm (for leaf temp. B)

c 10 g

d current

e 0°-50°C

g 0.5%

h -10°– 50°C
0 – 80% rh

i 30 m

j $10

4 100 V, 50 or 60 Hz

5

a Meter or multi-channel recorder

See page xii for Key

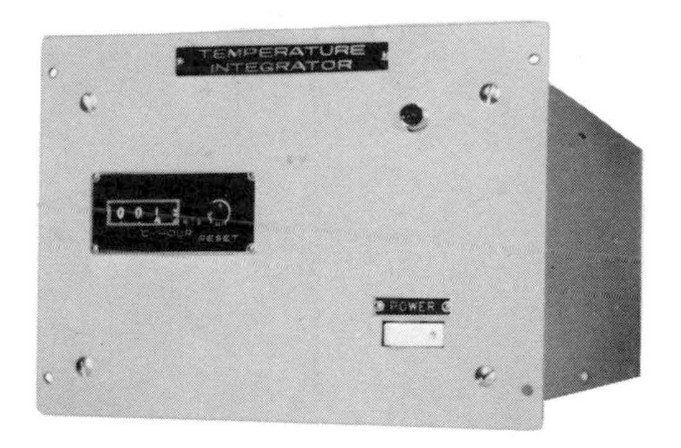

1 Temperature integrator

2 Iio Denki Ltd.,*
Yoyogi 2-27-18, Shibuya-ku,
Tokyo, Japan.

3

b 15 x 21 x 30 cm
c 7 kg
h -10°to + 50°C
0 to 80% rh
j $ 500

4 100V, 50 or 60 Hz

See page xii for Key

1 Aspirated psychrometer (3 PP 115)

2 Theodor Friedrichs,
2 Hamburg - Schenfeld 1,
Borgfelde 6, Postfach 1105,
Germany (BRD).

3
a Aspirated Pt resistance elements
b Housing 48 cm tall x 17 cm diam.
c 0.5 mg
d Change of resistance
g Depends on measuring bridge
h Above 0°C

4 24 or 220 V A.C.

6 Thermometers are exposed in double-walled chromium plated radiation shield and are aspirated at 4 m/sec by motor-driven fan.

See page xii for Key

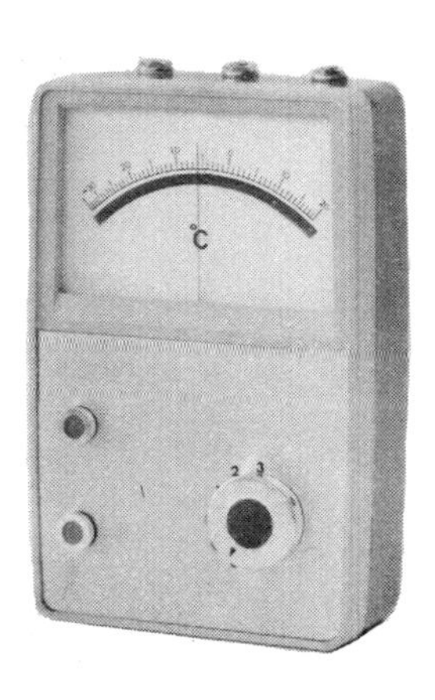

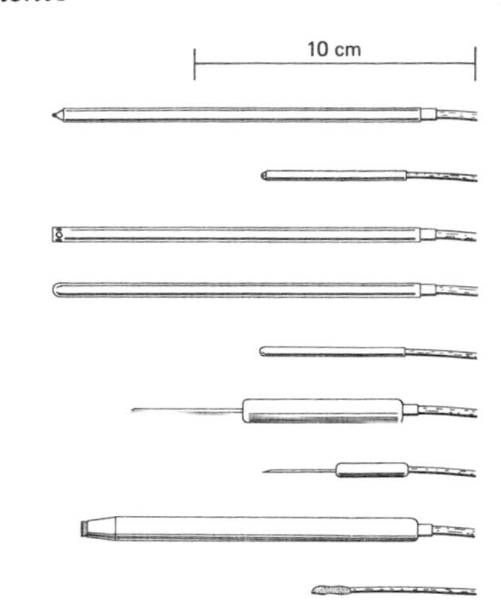

1 Thermistor thermometer (Model S)

2 Grant Instrument (Development) Ltd., Toft, Cambridge, U.K.

3
a Change of resistance
b Probes minimum diam. 1/8 in; meter 6 x 4 x ½ in
d Bridge voltage
e Any range of 10°, 20°, 50° or 100° C between -50° and 250° C
± 1% fs
As e
Standard 6 ft., extension available fitted with plug and socket
Meter £30
Probes £3 to £6

4 Internal battery

5
a Meter

6 Extra facilities include 2 ranges (switched) and provision for 8 channel switched input
A similar model is available for use with a wet and dry thermistor; reads rh directly in range 20 to 100%. (£50)
A third model measures temperature difference in range ±5° C
Wide range of matched probes available as shown above.

See page xii for Key

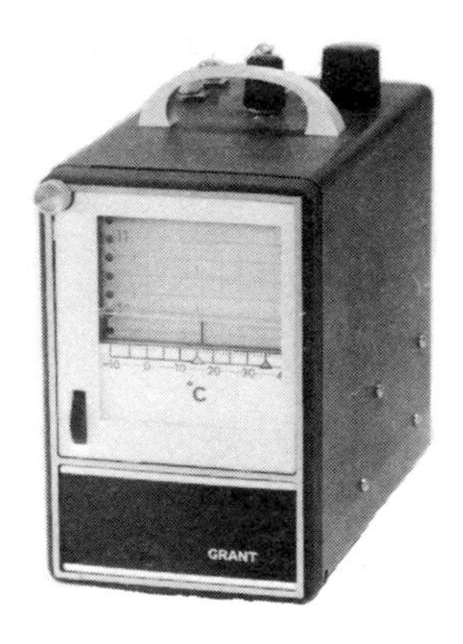

1 Temperature recorder

2 Grant Instrument (Development) Ltd.,
Toft, Cambridge, U.K.

3
a Thermistor and galvo with chopper bar
b 7 x 5 x 4 in
c 5lb
d Bridge voltage
e As required between 5°and 300°C
g $\pm$ 2% fs
j £130 plus probes

4 6V battery incorporated; one year life on
single channel

5
b Incorporated

6 Development of basic Rustrak recorder
Maximum 9 channels, quick scan every
¼, ½ or 1 hour. Chart width 2½ in;
choice of speeds from ¼ to 120 in/hr.
Wide range of sensors and waterproof box available.

See page xii for Key

1 Quartz crystal thermometer

Hewlett Packard Ltd.,*
224 Bath Road,
Slough, Bucks, U.K.

3

a Variation of frequency of quartz crystal with temperature
b Minimum $\frac{3}{8}$ in x $\frac{11}{16}$ in
c Minimum about ½ oz
d Digital
e -80°C to +250°C
f 2 sec
g Accuracy about ± 0.02°C but dependent upon temperature being measured and on linearity errors within that region. Resolution ± 10^{-4}°C or 10^{-5}°C
h -5°C to +55°C, rh. 95% at 40°C
i Standard at 12 ft but 1 mile or more with amplifiers which can be supplied.
j £1484 + duty at approx. £214 if applicable.

4 None

5

a Digital display on Nixie tubes but analogue outputs and digital printer outputs can be provided.
b Recorder output is variable from 1 mv fs upwards according to the range of temperature required. Fs of 0.01°C or optional 0.001°C are available.

6 Weight is 22 lb but portability limited by mains voltage requirement at either 230 V or 115 V.50Hz
Applications in oceanography, hydrology, measurement of heat output from leaves freezing points of animals and fish blood in pregnancy and environmental studies

7

a Professor Holliday
Dept. of Biology, Stirling University, Scotland.
b Dr. R. Tait,
Dept. of Oceanography, Liverpool University, (also N.I.O.)
c Dr. E.C. Coombe, Turner Dental Hospital, University of Manchester, Bridgeford Street, Manchester, U.K.

See page xii for Key

1 Thermistor thermometer

2 Laboratorio Ricerche Elettroniche,
Via Rutilia 19-18,
Milano, Italy.

3

a Change of resistance

b Minimum probe size 2 mm diam. x 50 mm long; meter 10 x 14 x 4 cm

d Bridge voltage

e -50° to +50°C

g ± 1.5% fs

j Probes £10 to £20, meter £14

4 Battery operated (3V)

5 Standard meter will accept range of probes for air, liquid and surface temperature measurement

See page xii for Key

Thermometer

Leanord*
3 Rue F. Baes,
59 Lille, France.

Change of resistance
5.2 x 40 mm
Voltage or digital signal
−35°to +45°C
± 0.1°C
Sensor −30°to +70°C
Interface 0°to 40°C
£1278

110/220V, 50Hz

Output in analogue form (0.1V /deg C)
or BCD

See page xii for Key

1 Thermistor thermometer

Light Laboratories,
10 Ship Street Gardens,
Brighton, BN1 1AJ,
Sussex, U.K.

3
- a Change of resistance
- b Probe, minimum diam. 0.7 mm x 22 cm long
- d Bridge voltage
- e As required
- g ± 1% fs
- j £60 for control unit and meter; £8 to £14 for probe depending on type.

4 Mains power or battery operated.

5 Meter incorporated.

6 A wide range of probe types is available, several designed specifically for medical use but with applications in plant studies.

See page xii for Key

1 Wet bulb and Globe Thermometer, Mark III, complete with anemometer probe, human oral probe and general purposes probe

2 Light Laboratories,*
10 Ship Street Gardens,
Brighton, BN1 1AJ,
Sussex, U.K.

3

a Thermistor in a DC Bridge
b Transit Case, 60 x 25 x 22 cm
Instrument: Rectangular, 25 x 14 x 15 cm
c Total 11kg
Instrument Weight: 3.5 kg
d Bridge voltage
e WB 15/70°C GL 15/70°C
DB 15/70°C WBGT Index 20/60°C
f All probes except globe 2 min: globe 10/15 min
g Resolution better than 1°C for air temp., 0.2°C for physiological temp.
h −10°to +70°C up to 100% rh
i 24 foot standard, maximum 100ft
j £235

5

a Moving coil microammeter, 200 microamperes at 350 ohms pivoted instrument, scale length 6 in, capable of feeding a high impedance recorder.

6 Measures wet, dry and globe temperatures, wet bulb globe temperature index (WBGT) and wind speed (0 to 500 ft min^{-1}) General purposes probe measures temperature in range −35°to 50°C. Instrument can be produced with several complete sensing leads selected by a switch. Powered by 2 9 - volt dry batteries.

7

a M.V.E.E. Dept. Prov., Mr. Ives, Chobham Lane, Chertsey, Surrey, U.K.
b Royal Navy, Sgn. Cmmdr. Walters, Physiological Lab., Alverstoke, Hants., U.K.
c Royal Aircraft Establishment, Dep. Aviation Medicine, Farnborough, Hants., U.K., Mr. D. Robertson.

See page xii for Key

1 Miniature Wet Bulb, Dry Bulb Globe Thermometer Index Instrument, Mk. 4.

2 Light Laboratories,*
10 Ship Street Gardens,
Brighton, BN1 1AJ,
Sussex, U.K.

3

b Rectangular, instrument 12 x 14 x 14 cm
Inst. and accessory box 30 x 12 x 14 cm

c 4 kg

j £350

6 This instrument is a miniature WGBT and is suitable for carrying in the field and measurements in small spaces. Specification as for Mark III except for weight and size

7

a M.V.E.E. Dept. Prov., Mr. Ives, Chobham Lane, Chertsey, Surrey, U.K.

b R.A.E., Dept. Human Engineering, Mr. Bolton, Farnborough, Hants, U.K.

See page xii for Key

1 Maximum-minimum thermometer (AM 17)

2 Mashpriborintorg,
Smolenskaya-Sennaya Square, 32/34,
Moscow G-200,
U.S.S.R.

3
a Expansion of toluene
b Recorder 35 cm diam. x 25 cm
c 5.5 kg
d Displacement of pen arm
e −30° to +10° C
h −35° to +35° C
i 2.5 m

6 The instrument is designed to record, on a blackened metal drum, the maximum and minimum temperatures close to the soil surface in spring. Expansion of toluene in the sensor displaces a recording arm attached to a manometric capsule by a helical spring.

See page xii for Key

1 Thermometer (M 54)

2 Mashpriborintorg,
Smolenskaya – Sennaya Square, 32/34,
Moscow G-200,
U.S.S.R.

3
a Change of resistance
b Meter and case 32 x 32 x 11 cm
c 46 kg
d bridge voltage
e −35° to +55 °C

5
a Meter incorporated

6 Intended for measuring soil temperature at 10 depths; instrument box includes selector switch, galvanometer and power supply.

See page xii for Key

1 Thermometer (AM 2 m)

2 Mashpriborintorg,
Smolenskay-Sennaya Square, 32/34,
Moscow G-200,
U.S.S.R.

3
a Change of resistance
b Meter 18 x 17 x 10cm
c Meter 1.8 kg, sensor & cable 0.8 kg
d Bridge voltage
e −30° to +45°C
i 2m standard

5
a Microammeter incorporated

See page xii for Key

1 Kata thermometer

2 Negretti & Zambra,
London, U.K.

3
a Cooling of thermometer with large bulb
b 23cm long
c 55g
d scale
e 95° to 100°F, 125° to 130°F, 145° to 150°F
f depends on wind speed and environmental temperature

6 For measuring low air velocity, only dry bulb readings are taken but for use as a 'comfort' meter, both wet and dry bulb readings are needed. A silk sleeve is provided for wet bulb measurement.
(See Bedford, 1964, p. 261)

See page xii for Key

1 Resistance thermometers (Stikons)

2 RdF Corporation,
23 Elm Avenue,
Hudson,
New Hampshire 03051, U.S.A.

3
a Change of resistance
b Thickness from 0.01 to 0.5 in
d Bridge voltage
f From 50 msec
h To 2000°F
j $ 6 to $ 40

5
a Several meters available for 100 ohm thermometers

6 A wide range of Stikon units is available designed for cementing to plane or curved surfaces.

See page xii for Key

1 Foil thermocouples

2 RdF Corporation,
23 Elm Avenue,
Hudson,
New Hampshire 03051, U.S.A.

3
a Thermo-electric effect
b Thin foil 0.0002 in and 0.0005 in thick
d Voltage
f 10 milliseconds
j $ 7 to $ 14

6 Other types of temperature, radiation and heat flux sensor available.

See page xii for Key

1 Temperature sensors (including wet-bulb thermometers)

2 Rosemount Engineering Co. Ltd.,*
Durban Road,
Bognor Regis,
Sussex, U.K.

3
a Change of resistance
b Minimum 0.15 in square x 0.05 in thick
d Bridge current; 0.4 ohms/°C
e −263°C to +800°C
f 0.1 sec minimum
g system accuracy 10^{-1} °C
system resolution 10^{-3} °C
h −200°C to +250°C
i 500 yd
j £3 to £30

4 Temperature transmitter or bridge system

5
a 1 mV/deg C signal or 4-20mA signal

7
a Dr. P.G. Jarvis,
Department of Botany,
University of Aberdeen,
St. Machar Drive,
Aberdeen, AB9 2UD, U.K.

See page xii for Key

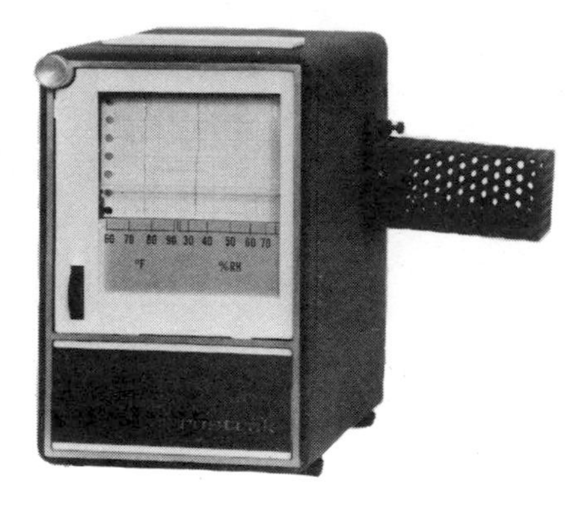

1 Temperature and humidity recorder

2 Rustrak Instruments Divn.,*
Manchester,
N.H. 03103, U.S.A.

3
b 6 x 4 x 5 in
c 3 lb
e 60° to 90°F; 10 to 90% rh
f 1 sec temp.; 30 sec rh
g $\pm$ 1°F, $\pm$ 4 % r h
i 200 ft
j $ 255

4 115 or 240V, 50/60 Hz

5
b Rustrak recorder 8 sec sampling period
½ in/hr

See page xii for Key

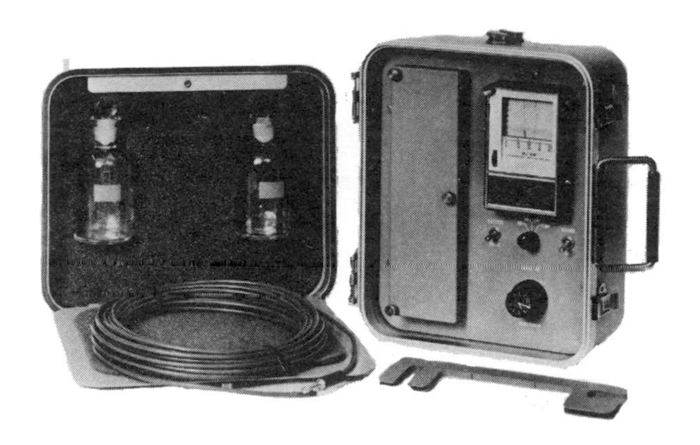

1 Temperature and dissolved oxygen meter (Model 192)

2 Rustrak Instruments Divn.,* Manchester, N.H. 03103, U.S.A.

3

b 9 x 11 x 14 in
c 24 lb
d Bridge voltage
e 0°to 40°C
g ± 2 % f s, ± 1°C
h −30°to +65°C
j $ 585 without oxygen probe; oxygen probe $155

4 12V Battery

6 One month unattended operation.

See page xii for Key

1 Thermistor psychrometer

2 Science Associates Inc.,
230 Nassau St.,
Box 320 Princeton,
New Jersey 08450,
U.S.A.

3

a Change of resistance
b 0.094 in diam. x 0.5 in long
c 2½ lb including aspirator motor and battery
d resistance
e −2°C to 50°C
g ± 0.25°C
j $322

4 None

5

a Complete portable system includes meter in carrying case, total weight 6 lb.

See page xii for Key

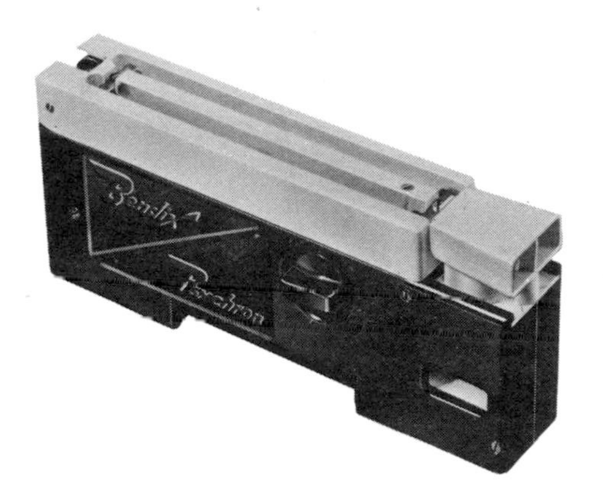

Psychrometer with mercury thermometer

Science Associates Inc.,
230 Nassau St.,
Box 320 Princeton,
New Jersey 08450,
U.S.A.

3½ lb including aspirator motor and battery
Thermometer scale
−15°to 45°C
0.1°C
$ 70

Built in illumination system for use in dark
and radiation shield for sunshine

See page xii for Key

1 Thermocouple

2 Spembly Technical Products Ltd.,*
Trinity Trading Estate,
Sittingbourne,
Kent, U.K.

3
a Thermo-electric effect
b 0.25 – 6.35 mm outside diameter
d voltage
e –200°C to 2000°C
f Dependent on type, size and environment
g As per B.S. 1041; can be calibrated
h As e
i 1500 feet
j On application

4 None

5
a Battery powered (portable) indicators for various thermocouple combinations, from –190°C to +1760°C

6 S.T.P. Ltd. manufacture thermocouples t suit specific requirements, all units being individually calibrated and tested as necessary. The small diameters available, and resulting low mass, makes these suita for a very wide range of applications in th high and low temperature fields, particul where reliability over an extended period of importance.

7
a U.K.A.E.A. Harwell,
Didcot,
Berks., U.K.
b Rolls Royce,
Derby, U.K.
c Edinburgh and S.E. Scotland,
Blood Transfusion Service,
Royal Infirmary,
Edinburgh, U.K.

See page xii for Key

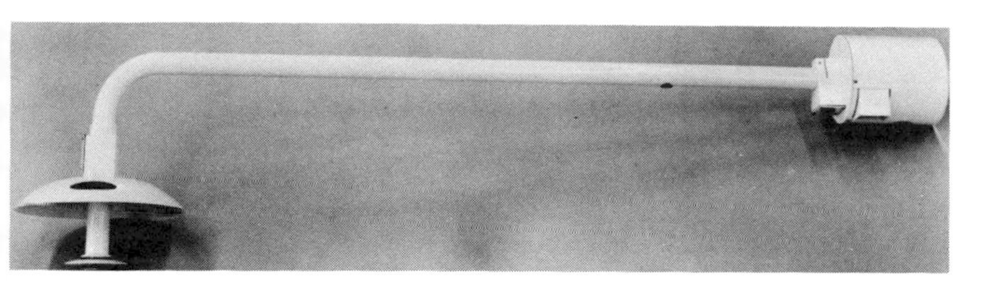

1 Aspirated thermometer shield

2 Teledyne Geotech,
3401 Shiloh Road, Garland,
Texas 75040, U.S.A.

3
a Radiation shield with aspirating motor
b Arm 60 in long, shield 11 in diam.
c 15 lb
g Radiation error less than 0.2°F at 1.6 cal cm^{-2} min^{-1}
h –40 to + 150 F
i –40°to +150°F
j $ 330

4 Temperature probe available as extra,
$70. Aspirator motor needs power supply
115 V, 60Hz

See page xii for Key

1 Temperature integrator (SAMI/T)

2 TEM Sales Ltd.,
Gatwick Road,
Crawley, Sussex, U.K.

3
a Electroplating in miniature electrolytic cell integrates bridge current in thermistor circuit
b 7 x 4 x 2 cm
c 78 g
e −20° to 55°C
f approx 1 min
g Within ± 0.5°C over most of range
i 1 m
j Complete integrator ca £80; replay machine £110

4 Replay machine

5
a Digital readout on replay machine

6 SAMI (socially acceptable monitoring instrument) is designed to be carried unobtrusively by human subjects. Can b set to give true mean temperature or me above or below a threshold. A similar instrument can be used to record total heart beat. (See Humphrey and Wolff, 1967, p. 241)

See page xii for Ke

1 Thermocouple

2 T.F.D.L., Dr. S.L. Mansholtlaan 12, Wageningen, Netherlands.*

3

a Thermo-electric effect
b Wire diam. from 0.010 to 1.0 mm
c Dependent on dimensions
d voltage
e Usually −30°to +50° C
f 1 to 60 s in stagnant air, depending on wire diameter
g Accuracy ± 0.05°C (−30°to +50°C), resolution depends on the method of mV-measurement
h No essential restrictions for ambient conditions
i Not limited
j Price depends largely on type and length of thermocouple wire

4 Reference temperature, e.g. ice-point

5 mV meter or recorder

6 Copper-constantan and other thermocouple materials; mounted under double radiation shield for air temperature and wet-bulb temperature; mounted in hypodermic needle for soil temperature

7 Agricultural University, Wageningen, Netherlands.

See page xii for Key

1 Semi-conductor temperature sensor

2 T.F.D.L., Dr. S.L. Mansholtlaan 12, Wageningen, Netherlands.*

3
a Emitter-base voltage variation with temperature at constant collector current
b 2.5 mm diam. sphere
c 2 g
s Voltage
e Used in the range –30° to +80° C
f 15 s
g ± 0.05° C, resolution 0.02° C
h –15° to +70° C, not affected by humidity
i Cable length not limited
j £4; electronic circuit with meter in wooden box, £105.

4 Battery powered electronic control circuit

5
a The linear scale of the moving coil meter is directly calibrated in degrees Celsius, e.g. 100 micro-amp.meter for 10° C full scale range
b The electronic circuit has a 10 mV, 200 μ recorder output for a potentiometric or a galvanometric recorder

6 Portable, for use in field and laboratory. The sensor is in the point of a stainless st needle for measuring soil temperatures in field at greater depths

7 Agricultural University, Wageningen, Netherlands.

See page xii for Key

1 Thermocouples

2 Thermoelectric Co. Inc.,*
Saddle Brook,
New Jersey 07662,
U.S.A.

3

a Thermo-electric effect
b minimum 1/25 in diameter
c Dependent on type
d Voltage
e −250°C to 2000°C
f 0.1 sec for 1/16 in diameter
i Dependent on characteristics of recorder
[illegible] Minimum £4

6 A complex range of equipment is available, including probes, connectors, wires and instruments.

See page xii for Key

1 Thermistor probe

2 Yellow Springs Instrument Co., Inc.,*
P.O. Box 279,
Yellow Springs, Ohio 45387, U.S.A.

3
a Resistance change
b Various from 1/16 in diameter
c From 10 mg to approx 100 g with lead
d Resistance change; interchangeable but not linear 400 Series
e −80° to +150°C
f 0.1 sec to 7 sec depending on mounting and medium
g Changes of up to hundreds of ohms per degree Centigrade
h Instrument 0° to 50°C probes −80° to 150°C to 100% rh
i Unlimited
j $5 to $50 depending on mounting

4 Resistance bridge

5
a YSI offers many meter models to give dir[e]
temperature readout. Meters mainly 50 microamp, 2000 ohm input impedance.
b Resistance recorder, or YSI meter readou[t]
have 100 mV recorder outputs

6 YSI battery powered thermistor thermom[eters]
cover the ambient range in many sub-ranges maximizing accuracy and readabili[ty]
Useful for soil, surface, liquid and air tem[p]eratures. Linear output probes allow utili zation of direct differential readings.

7 In very wide use in the United States

See page xii for Key

1 Thermistor probe

2 Yellow Springs Instrument Co. Inc.,*
P.O. Box 279,
Yellow Springs, Ohio 45387, U.S.A.

3
a Resistance change
b 2.8 mm x 3.7 mm semi-cylinder plus probe forms
c 20 mg to approx. 50 g in probe form
d Voltage or resistance linear with temperature
e −55°C to +100°C in a number of ranges
f 1 sec in well stirred oil

g Interchangeability ± 0.15 °C; linearity 2 parts/1000 over range
h 0°– 50°C for instruments provided
i 3 m standard; essentially unlimited
j $10 to $34 depending on mounting

4 YSI provide signal conditioner

5
a Digital voltmeter. Need 500k ohms or more. DVM reads directly in temperature including sign.

6 Can be hooked up to read differential temperature

7 In wide use in the United States

See page xii for Key

1 Temperature integrator

2 Yellow Springs Instrument Co., Inc.,*
P.O. Box 279,
Yellow Springs, Ohio 45387, U.S.A.

3

a Coulometer in Wheatstone bridge
b As thermistor probes (p. 41)
c As thermistor probes
d Resistance
e −29°to 49 °C
g Difference between set point and probe readable to + 0.025°C hours
h −20°to +50° C
i 3 m standard – essentially unlimited
j $ 345 instrument, plus $ 20 ro $34 for probe

5

a Archimedes spiral for reading of coulometer movement

6 The instrument is battery powered and portable and indicates:

(1) mean temperature and degree hours or;
(2) mean temperature and degree hours diffe
between two points or;
(3) mean temperature and degree hours abov
or below a set temperature threshold

7 In wide use in the United States

See page xii for Key

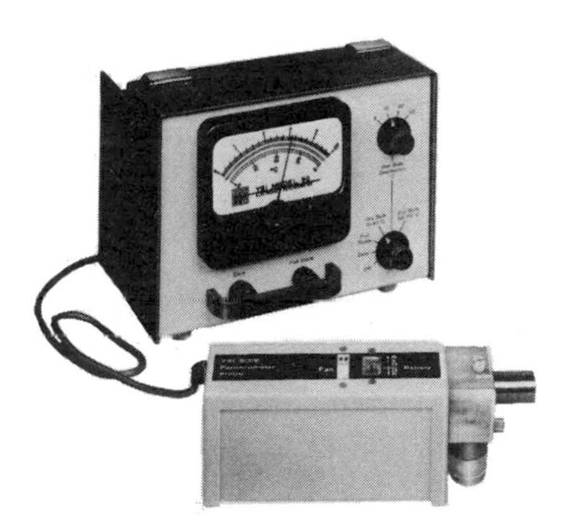

1 Thermistor psychrometer (YSI90)

2 Yellow Springs Instrument Co., Inc.,*
P.O. Box 279,
Yellow Springs, Ohio 45387, U.S.A.

3
a Change of resistance; direct reading of wet and dry bulb differential
b probe 23 x 9 x 5 cm; meter 23 x 18 x 15 cm
c Approximately 0.5 kg
d Resistance
e Temperature 0°– 70°C
Wet bulb depression: 0°–5°C, 0°–10°C, 0°–40°C
f 2 min
g Wet bulb depression resolvable to ± 0.02°C accuracy better than 2% when ambient is 0°–35°C and above, and rh is above 50. Better than 3.5% – when ambient is 0°–35°C and rh is less than 50%
h Instrument –20° to 50°C
Probe 0° to 70°C
Humidity – full range
1.37 meters standard – limit to 8 meters
$475

4 YSI model 90 readout with 9019 probe

5
a Direct readout of differential and dry bulb

6 Battery powered, portable, electrically aspirated radiation protected sensors

7 In wide use in the United States

See page xii for Key

1 Temperature-radiation shield (aspirated)

2 R.M. Young Company,*
42 Enterprise Drive,
Ann Arbor, Michigan 48103, U.S.A.

3
a Double wall evacuated glass tube (Gill)
b Horizontal Arm 46 in long with shield at end 18 in high
c 7½ lb
g Maintains sensor within 0.05°C of ambient temperature
h −30°C to +70°C, 0–100 rh
j $196

6 Designed for tower mounting. Blower requires AC power (7 watts) – also available for DC on special order. Shield is designed particularly for temperature gradient measurements.

7
a Department of Meteorology & Oceanography, University of Michigan, Ann Arbor, Michigan, U.S.A.
b Canada Center for Inland Waters, Burlington, Ontario, Canada.
c U.S. Weather Bureau, Silver Spring, Maryland, U.S.A.

See page xii for Ke

2
Hygrometers and Infra-Red Analysers

Hygrometers are sensors for measuring the water vapour content of a sample of air. Commercial hygrometers are based on several principles: condensation of vapour on a surface cooled to the dew-point; equilibrium temperature of lithium chloride in the solid and liquid phases; capacity of an aluminium oxide layer; resistance of a sulphonated polystyrene layer. Hygrometer sensors all suffer to a greater or lesser extent from hysteresis and several types are permanently damaged when they are exposed to saturated air. The lithium choride 'dewcell' is the type most commonly used in micro-meteorological studies.

Infra-red analysers can be used to measure either the carbon dioxide or the water vapour content of an air sample. Most infra-red analysers depend on the differential absorption of radiation in two sample tubes with identical geometry so the technique is well suited to measuring small vertical differences of concentration in and above vegetation. Maximum precision is about 0.1 vpm CO_2 and 0.01 mb water vapour pressure. Absolute CO_2 analysers can cover the normal atmospheric range 200 to 400 vpm, or the range that is appropriate for respiratory studies of animals and man, about 5 to 10%. Conventional infra-red analysers incorporated a valve amplifier but more advanced designs using solid state electronics are now becoming available.

1 Infra-Red Gas Analyser.

2 The Analytical Development Co., Ltd.,*
Salisbury Road,
Hoddesdon, Herts., U.K.

3

a Gas filled detector absorbing infra-red radiation
b 30 x 50 x 38 cm or for rack mounting
c Approx. 70 lb (32kg)
d Current (10mA); voltage (10 V)
e Carbon Dioxide 0 – 5000 ppm absolute, 0 – 50 differential †
f Typically 7 sec, for 90% of final reading
g Accuracy of directly calibrated meter scale better than 2%; repeatability 1%.
h 0°to 40° C
i Any length of cable may be used for remote readout.
j £600

4 Gas sampling system designed to meet t needs of the application

5 As required, may be by meter, indicator recorder, numerical indication, B.C.D. fo computer feed or print out

† Water vapour minimum 0.2%, maximum 5.0% by volume full scale (2-50 mb); absolute or differential measurement.

See page xii for Key

Dew point hygrometer

Analytical Instruments,
Fowlmere, Royston,
Herts, SG8 7QS, U.K.

attention by dew of radiation from
Nickel 63 source
28 x 13 x 24 cm deep
4 kg
Voltage
−60°to +40° C dewpoint
Cooling rate 6° C/min to −10° C
3° C/min to −60° C
± 1°C over whole range
0°to 40° C for quoted accuracy
Manual version £290; automatic version
£520

110 or 230 V, 50/60 Hz, CO_2 cylinder for
coolant

Incorporated

See page xii for Key

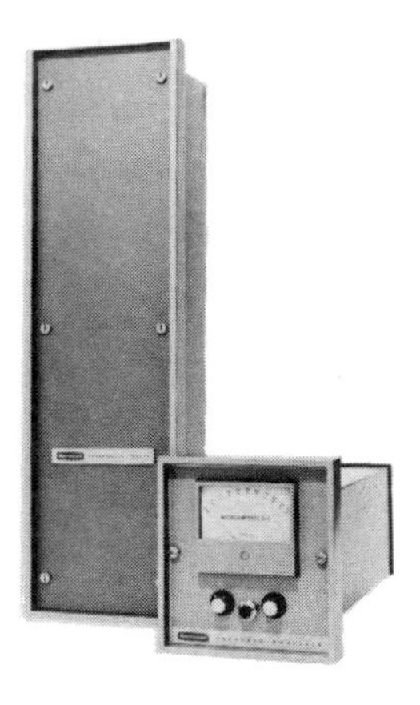

1 CO_2 Infra-Red Analyser

2 Beckman Instruments,*
2500 Harbor Boulevard,
Fullerton, California 92634, U.S.A.

3

a Differential absorption of radiation
b 30cm high x 72cm long x 24cm wide
c 34kg
d analog: 0-10mv, 0-100mv, 0-1 Volt, 0-5 Volt or optional 1-5ma, 4-20ma, 10-50ma
e 0-350 vpm CO_2 standard, other CO_2 ranges up to 100%; also other infrared absorbing gases
f 90% response in 0.5 sec
g 0-350 vpm CO_2 $\pm$ 1% of full scale; resolution 1 vpm
h +16°to 41° C for Model 215-A; for Model 315-A, −29°to 49°C (−20°to +120° F) relative humidity to 100%
i Model 215-A is self-contained, no cable. Model 315-A up to 152 meters (500ft)
j $2860, Model 215-A

4 Calibration sample representative of 80% of the CO_2 up scale range and CO_2 – scrubbed N_2 for zero gas

5

a available as standard
b voltage as stated in Item 3 (d) above; for current output maximum permissible load is 8000 ohms for 1-5ma, 2000 ohms 4-20ma and 700 ohms for 0-50ma.

6 Model 215-A is designed for laboratory with amplifier, analyser and meter in one box for bench top use; the 315-A for remote mounting of the analyser section and local mounting of the amplifier meter.

7

a Research Centre of France, Veir-le-Petit, France.
b University of Hawaii, Environmental He[alth] Dept., Honolulu, Hawaii, U.S.A.
c Bavarian Biological Research Institute, Munich, Germany (BRD).

See page xii for Key

1 Thermo-electric dew point hygrometer

2 Cambridge Systems Inc.,
50 Hunt Street,
Newton, Massachusetts 02158, U.S.A.

3
a Detection of condensation on mirror by phototransistor
b 9 x7 x 12 in
c 12 lb
d Meter reading or 0 to 50 mV
e –40°to 50° C dewpoint
f 2° C per sec maximum
g about ± 1° C over normal range
h –40°to 50° C
i 100 ft

4 50 W 115V 50-70Hz or 230V

5
a Meter scale reads –40°to 50° C

6 The dew point sensor is available in a number of different instruments.

See page xii for Key

1 Hygrometer (EH-505)

2 Iio Denki Ltd.,
Yoyogi 2-27-18, Shibuya-ku,
Tokyo , Japan.

3
b 15 x 21 x 30 cm
c 6.5 kg
e 20 to 100 % rh
g ± 2%
h 5°to 45°C
0 to 80 % rh
j $500

See page xii for Key

1 Hygrometer

2 Foxboro Co.,
Foxboro, Mass, U.S.A.

3
a Equilibrium temperature of lithium chloride
d Bridge voltage
e −50° to + 142°F dew-point
g ± 1°F
£200 (including dry bulb)

See page xii for Key

1 CO_2 Infra-Red Gas Analyser Model SB2

2 Sir Howard Grubb Parsons & Co. Ltd.,*
Shields Road, Walkergate,
Newcastle-upon-Tyne,
NE6 2YB, U.K.

3
a Differential absorption of radiation
b 0.81 x 0.43 x 0.36 m
c 68 kg
d Voltage or current
e vpm levels most infra-red absorbing gases, e.g. 0-500 vpm H_2O or CO_2
f 20 sec approx.
g ± 2% fs accuracy
± 1% fs reproducibility
h max 40°C 100% rh
i Limited by losses in cable only
j £800

4 Optional recorder etc.

5
a Standard built-in meter
b Millivolt recorder any range between 0-10 mV and 0-500 mV
Milliamp recorder – 0-1 mA

6 A versatile general purpose instrument with high sensitivity and stability, single and multi-range instruments available

See page xii for Key

1 CO_2 Infra-Red Gas Analyser (MGA2)

2 Sir Howard Grubb Parsons & Co. Ltd.,
Shields Road, Walkergate,
Newcastle-upon-Tyne,
NE6 2YB, U.K.

3
a Absorption of infra-red radiation
b 37 x 51 x 15 cm
c 22 kg
d 0.1-1 mA, external resistance 10 k ohms
e Most sensitive range 50 vpm fs
f Half-time 0.6 sec
g 2% fs

4 Power supply 110 or 220 V
50Hz or 110 V, 60Hz

5
a Meter incorporated
b Recorder, see 3 d

6 A new single tube design with fully transistorized circuitry. Compensated for temperature changes.

See page xii for Key

1 CO_2 Infra-Red Analyser (URAS 2)

2 Hartmann & Braun,
6 Frankfurt 90,
Postfach 900507,
Germany (BRD).

3
a Differential absorption of radiation
b 48 x 44 x 23 cm
c 35 kg
d 0-20 mA, max resistance, 750 ohms; or 0-5 mV
e maximum sensitivity 0-50 vpm CO_2
f 90% response in 0.5 sec
g Drift less than 2% fs per week
h 10° to 45° C

4 220V ± 15%, 50Hz

5
a Incorporated
b Potentiometric

See page xii for Ke

1 Lithium chloride hygrometer (15 - 7012)

2 Hygrodynamics, Inc.,*
949 Selim Road,
Silver Spring, Maryland 20910, U.S.A.

3
a Equilibrium temperature of lithium chloride
b 2 x 2 x 5 in
c 10 oz
d 0-5V DC; or 0-150μA
e 10-99% rh
f 10 sec
g ± 3% rh accuracy 0.5% rh resolution
h 20°-120° F
i 1,000 ft
j $484

4 8 to 32V, DC, power supply 30 mA.

5
a 0-100μA meter
b Potentiometer (200 k ohms)

7
a U.S. Forestry Service.
b U.S. Department of Agriculture.
c University of California, U.S.A.

See page xii for Key

1 Lithium chloride hygrometer

2 Hygrodynamics, Inc.,*
949 Selim Road,
Silver Spring, Maryland 20910, U.S.A.

3
a Equilibrium temperature of lithium chloride
b Cylindrical 2 in x 3/8 in diam.
c 1 oz
d change of resistance
e approx. 12% rh each sensor – 8 sensors required to cover 5 to 99% rh
f 2 to 3 sec
g ± 1.5% rh accuracy; 0.15% rh sensitivity
h –60° to +100° F
5 to 99% rh
i 200 ft with indicator
500 ft with recorder
j $600 complete

4 Indicator battery or A.C. operated

6 Either battery or line operated unit is portable

7
a U.S. Forestry Service.
b U.S. Department of Agriculture.
c University of California, U.S.A.

See page xii for Key

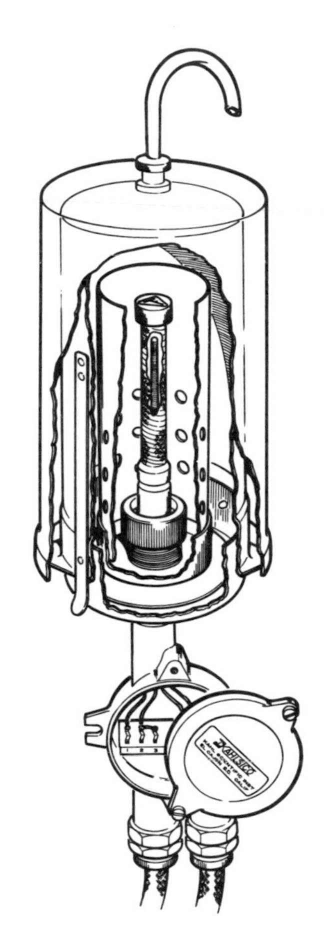

1 Lithium chloride hygrometer

2 Kahl Scientific Instrument Corp.,
P.O. Box 1166,
El Cajon, California 92022, U.S.A.

3
a Equilibrium temperature of lithium chloride

b 12 cm diam. x 32 cm in self-aspirating shelter
c 3lb complete with shelter
d resistance of Pt resistor
f ca 30 sec
g ca $\pm$ 0.3°C

4 24 volts, 50 to 60 Hz

5
a resistance meter

See page xii for Key

1 CO_2 Infra-Red Analyser

2 Mine Safety Appliances Co.,*
Braddock, Thomas and Meade Streets,
Pittsburgh, Pennsylvania, U.S.A.

3
a Differential absorption of radiation
b 19 x 12 x 13 in
c 76 lb
d 0-10 to 1-100 mV
e Maximum sensitivity fs 0 - 30 vpm CO_2
standard sensitivity 0 - 150 vpm CO_2
f 90% response in 5 sec
g $\pm$ 1% fs
h 0°to 50° C
j standard model $2500
sensitive model $3500

5
a incorporated
b potentiometric

See page xii for Key

1 Aluminium oxide capacitor

2 Moisture Control & Measurement Ltd.,
Thorp Arch Trading Estate,
Boston Spa, Yorkshire LS23 7BJ, U.K.

3
a Capacitance change
d 0–50 mV, 1 k ohm
e −100°to +100° C dewpoint
f instantaneous
g ± 3% fs; reproducible to 0.5%; resolution 0.1% rh or 0.2° C
i Low capacitance cable needed (17pf/ft)
j From £180

4 110V or 240V 40-60 Hz (stabilized)

5
a Supplied
b Potentiometer 0–50 mV, to 1 k ohm

See page xii for Key

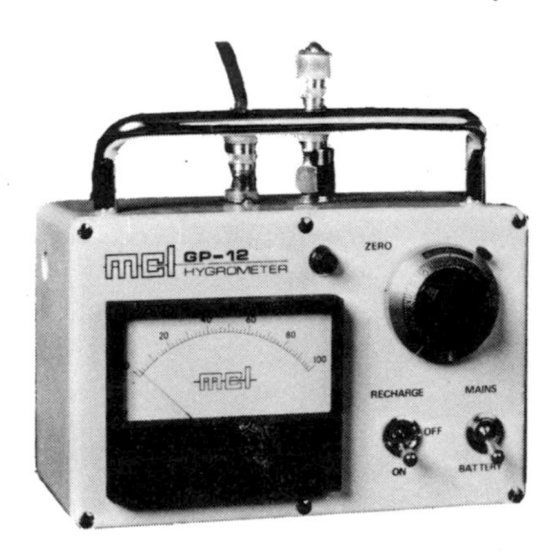

1 Hygrometer(GP 10)

2 Molecular Controls Ltd.,
Scottish Life House,
Leeds, 1, Yorkshire, U.K.

3
a capacitance of aluminium oxide element
b element 2 cm diam. x 10 cm long:
case 10 x 16 x 19 cm
d bridge voltage
e −80°C to +50° dewpoint
f 2 sec
g ± 1% fs; 0.2°C dewpoint
i 1 m standard, extension possible
j £165

4 110 or 240 V, 40 to 60 Hz

5
a Incorporated
b Output 0 - 50 mV provided

6 Battery operated version available (GP 11)
Element undamaged by high rh or wetting

7
a Central Electricity Generating Board,
Trawsfyndd Power Station,
Trawsfyndd,
Merionethshire, U.K.

b W.C. Holmes & Co. Ltd.,
Turnbridge,
Huddersfield, Yorkshire, U.K.

See page xii for Key

Hygrometer (PCRC 55 and 11)

Phys – Chemical Research Corp.,
36 East 20th Street,
New York, N.Y. 10003, U.S.A.

Resistance of sulphonated polystyrene film
Minimum 2.2 x 4.1 x 0.2 cm
Resistance change
0 to 100% rh
30 sec
Hysteresis about ± 2.5% rh
–60°to 200° F
£10 (unmounted)

Resistance bridge (10 M ohms); power
supply 1 mA, 20 Hz

Can be supplied mounted in probe or
unmounted

See page xii for Key

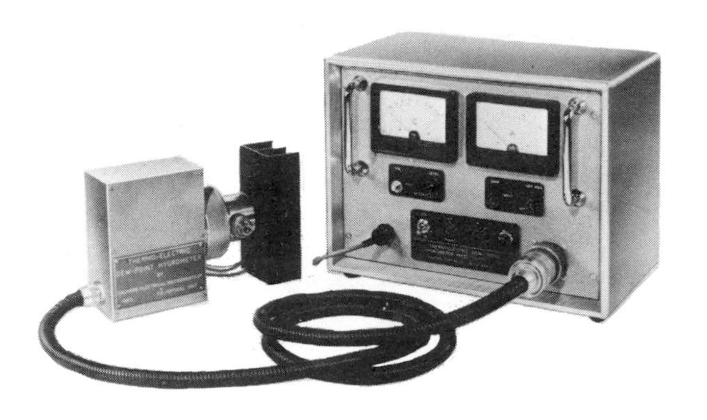

1 Dew-Point Hygrometer (Mk 3)

2 Salford Electrical Instruments Ltd.,*
Peel Works,
Barton Lane,
Eccles, Lancs., U.K.

3
a Detection of dew on surface cooled thermo-electrically
b Sensing Unit (Thermo-electric/Optical Unit) 24 x 15 x 8 cm
c Sensing unit 2.7 kg
d Voltage
e + 40 °C to – 40 °C Dew-Point
f With the sensor operating in a 20° C ambient the instrument will attain a dew-point of –10° C within 1 min of switch on and a dew-point of –30° C within 3 min
g Standard output from Control Unit ± 1½ C. Thermocouple output ± ½ °C
h –20° to +50° C, 95% rh
i Standard 6ft
Special 100ft
j Standard – Control Unit + Sensing Unit £357

4 Control Unit Size - 34 x 18 x 26 cm
Weight-11.6 kg
Operates from 240 or 115 volts, 50 Hz

5
a Meter incorporated
b Adjustable output provided for millivolt recorder

6 Correct measurements in still or moving air

7
a Ministry of Technology,
National Gas Turbine Establishment,
Pyestock,
Hampshire, U.K.
b Agricultural Research Council,
Meat Research Institute,
Langford,
Bristol, BS18 7DY, U.K.

See page xii for Key

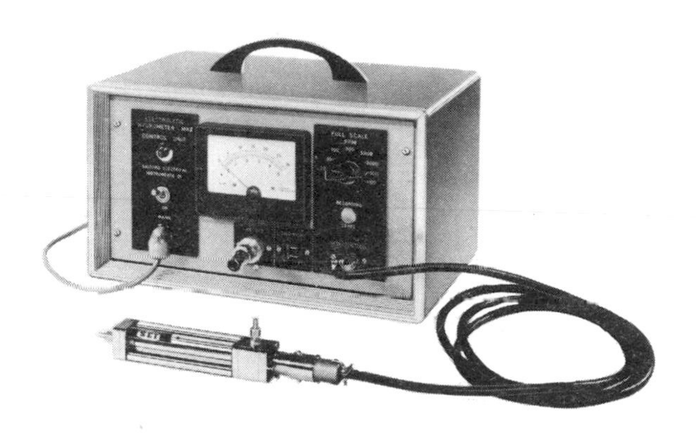

1 Electrolytic Hygrometer

2 Salford Electrical Instruments Ltd.,*
Peel Works,
Barton Lane,
Eccles, Lancs., U.K.

3
- a Continuous electrolysis of water in gas stream
- b Sensor (Detector Cell) – 18 x 3.5 x 2.5 cm
- c Sensor (Detector Cell) – 350 g
- d 0-30,000 vpm with a gas flow of 10 ml/min
- f Approx. 2 minutes for 90% response
- g Accuracy + 3%
- h 0°- 40°C, 95% rh
- i Unlimited – cable loop resistance must be less that 50 ohm.
- j Detector Cell £74

4 Control Unit. Size – 31 x 18.5 x 20 cm
weight – 5.3 kg
cost – £118

5
- a Meter incorporated
- b Adjustable output available for millivolt recorder

6 Continuously measures gas wetness only with a constant known gas flow rate. Measures moisture content in discrete quantities of gases, liquids or solids; able to detect less 0.1μg of water.

7
- a Central Electricity Research Laboratory,
Kelvin Avenue,
Leatherhead,
Surrey, U.K.
- b Pirelli General Cableworks Ltd.,
Leigh Road,
Eastleigh,
Hampshire, U.K.

See page xii for Key

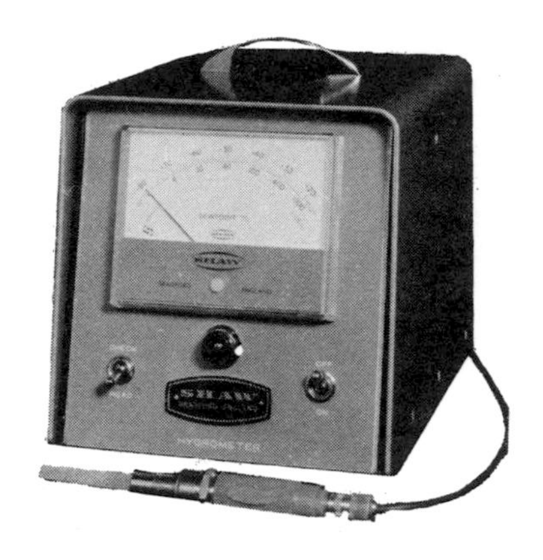

1 Humidity meter

2 Shaw Moisture Meters,
Rawson Road, Westgate,
Bradford, Yorkshire, U.K.

3
a Change of capacitance
c Smallest available sensor 10 g
d Voltage
e 0 to 98% rh
f Less than 1 sec
g ± 3% fs
i 100yd
j £220

4 Constant temperature unit available for
field measurements (£100)

5
a Meter incorporated
b 0 – 25 mV output for potentiometer

See page xii for Ke

Hygrometer (PJT)

Sina Ltd.,
8050 Zurich, Switzerland.

Semi-conductor sensor
Meter 19 x 12 x 7 cm
Meter 1 kg
Bridge voltage
15 to 95% rh
$<$ 10 sec
Accuracy better than 3% for 30 to 80% rh
repeatability better than ± 2%
0° to 40° C

Meter incorporated

A more accurate hygrometer (NJT) is also available, repeatable to 0.5% in range 5 to 95% rh and -10° to 40° C. This instrument has a servo-operated AC bridge and is fully transistorized.

See page xii for Key

1 Semi-conductor hygrometer

2 T.F.D.L., Dr. S.L. Mansholtlaan 12, Wageningen, Netherlands.*

3

a Wet and dry bulbs
b 2.5 mm sphere without cotton
c 2 g
d Voltage
e 30-100% rh
f 30 sec
g Accuracy 2% rh; resolution 1% rh
h 0° - 40° C
i Cable length not limited
j £47; electronic circuit with meter in wooden box, £110

4 Battery operated, the wet bulb must be ventilated with min. air velocity 1.5 m/s

5

a Temperature reading direct in °C from moving coil meter scale, rh reading with Mollier diagram
b Electronic circuit has recorder output fo[illegible] potentiometer recorder 10 mV or galva[illegible] meter recorder 200 micro-amp.

6 Portable, used for rh calibrating purpose[illegible] in the laboratory and rh determinations in the field.

7

a Agricultural University, Wageningen, Netherlands.

See page xii for Key

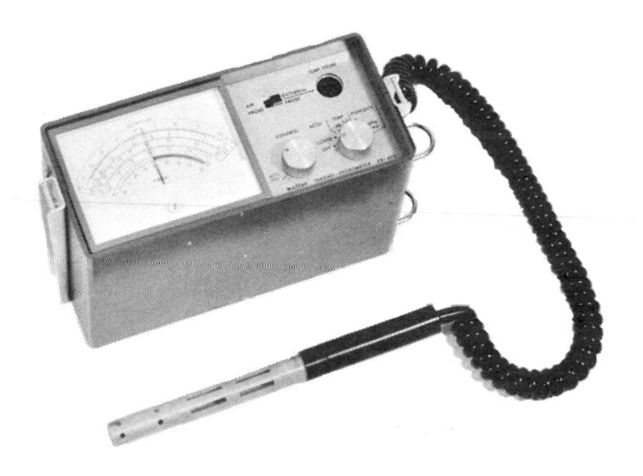

Hygrometer with thermometer (EP 400)

Wallac Oy,*
Turku 5,
Finland.

Equilibrium temperature of lithium chloride
Probe ca. 1 cm diam., instrument 19 x 12 x 9 cm
3 kg
Voltage
Dewpoint -20°to +60° C, 15 to 90% rh
temperature -20°to +120° C
1 sec
± 2% rh at 50% rh; ± 1° C
– 20°to 45° C
2 m
£135

Uses built-in nickel-cadmium rechargeable accumulators

Built in meter calibrated for humidity and temperature recorder available ; recorder output 4.4V, 0.5 mA

6 Interchangeable probes available for special applications, miniature recorder available.

See page xii for Key

1 Lithium chloride hygrometer (YSI 91)

2 Yellow Springs Instrument Co. Inc.,*
P.O. Box 279,
Yellow Springs, Ohio 45387, U.S.A.

3

a Equilibrium temperature of lithium chloride

b Cylindrical; 1 cm diam. x 6 cm long

c 100 g

d Resistance change at thermistor

e Dew point -12°C to +42°C;
Vapour pressure 1.8 to 61.5 mmHg

f 1°C/min after warm up

g Meter accuracy better than + 0.9°C,
Recorder ± 0.4°C; read 0.25°C dew point,
1 mm vapour pressure

h 0° to 50°C
0 to 100% rh

i 3 meters standard – use to 33 meters

j $275 + $70

4 Indicator (YSI 91 HC), 3kg

5

a As specified

b 0 to 10 mv output for ambient and dew-point
Any 50 k ohm or more input impedance recorder

6 Line operated (230V available) – replac bobbins
Unique adiabatic cell available – very s sensor

7 In wide use in the United States

See page xii for Key

1 Lithium chloride hygrometer

2 Yokogawa Elec. Co.,*
Nakamachi 2-9-32, Musashino City,
Tokyo, Japan.

3
a Equilibrium temperature of lithium chloride
b 34 x 2.86 cm
c 800 g
d Voltage
e 0 to 10 mV
f 40 sec
g 0.5%
h Wind velocity is lower than 20 cm/sec
i 50 m
j $125

4 Dewcell transducer and power supply
(100V, 50 Hz)

5
a Recorder (-45°to -60°C)

See page xii for Key

3

Solarimeters

Until 1960 only a handful of solarimeters was available commercially but in the last decade the number on the market has increased by an order of magnitude, largely in response to a demand from micrometeorology. According to the terminology recognized by the World Meteorological Organization, an instrument for measuring total (direct and diffuse) radiation on a horizontal surface is a **pyranometer** but the term **solarimeter** is commonly used. A **pyrheliometer** is an instrument for measuring the direct component of solar radiation on a surface at right angles to the solar beam. A pyranometer is usually a thermopile exposed under a glass hemisphere whereas the thermopile in a pyrheliometer is exposed at the end of a tube pointing at the sun.

Thermopile instruments contain a large number of thermocouple junctions in series. The 'hot' junctions of the thermopile are always painted black and the 'cold' junctions are either painted white or attached to a heat sink. Thermopile instruments have the advantage of a uniform sensitivity over the whole solar spectrum. **Silicon solar cells** give a much larger output than thermopiles but their non-uniform spectral response is a disadvantage in some measurements.

Thermopiles are now available in the form of **tube solarimeters** for measuring the average radiation in a crop canopy. As they are inherently less accurate than conventional dome solarimeters they should be used for measurements of relative rather than absolute radiation.

The output from any type of solarimeter can be integrated with the systems described on p. 225 *ff* but several firms are now marketing a solar cell, integrator and counter in one convenient unit.

1 Bimetallic actinograph

2 C. F. Casella & Co. Ltd.,
Regent House,
Brittania Walk,
London, N.1., U.K.

3

a Expansion of black bimetallic element
b 34 x 18 x 28 cm
c 13 kg
d Pen record on chart
e Appropriate for solar radiation
f Slow

i None

4 None

5

a Paper chart mounted on clockwork driven drum

6 British Meteorological Office Pattern Mark III

See page xii for Key

1 Tube pyranometer

2 EKO Instruments Trading Co. Ltd.,*
2-1 Ohtemachi 2-chome,
Chiyoda-ku, Tokyo 100, Japan.

3
a Thermopile
b 1 x 20 cm
c 0.4 kg
d Voltage
e 0–20 mV
f 99% response in 55 sec
i 50 m
j $200

4 None

5
a Millivolt meter
b Potentometric recorder

6 The instrument may be used in crop canopies and below trees.

7
a Dr. Uemura Kenji, Div. of Meteorology, National Institute of Agricultural Sciences, 2-1 Nishigawara, Kita-ku, Tokyo 114, Japan.
b Institute of Agricultural Electricity, Central Institute of Electric Power, Abiko-machi, Higashikatsushika-gun, Chiba prof., Japan.

See page xii for Key

1 Pyranometer

2 EKO Instruments Trading Co. Ltd.,*
2-1 Ohtemachi 2-chome,
Chiyoda-ku, Tokyo 100, Japan.

3
a Moll-type thermopile
c 2 kg
d Voltage
e 0 -20 mV
f 99% response in 7-8 sec
i 20 m
j $210

4 None

5
a Millivoltmeter
b Millivolt recorder

7
a Lab. of Physics in Upper Air Layer,
Institute of Meteorology, Kita 4-35-8,
Koenji, Suginami-ku, Tokyo, Japan.

b Ist Lab. of Physics, Div. of Meteorology
National Institute of Agricultural Scienc
2-1 Nishigahara, Kita-ku, Tokyo 114,
Japan.

See page xii for Key

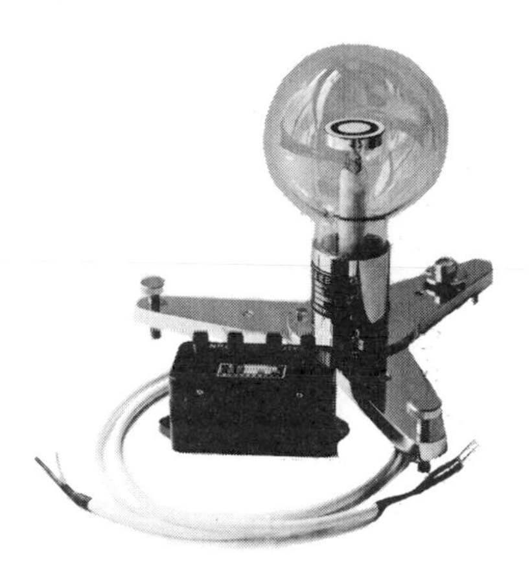

1 Pyranometer

2 EKO Instruments Trading Co. Ltd.,*
2-1 Ohtemachi 2-chome,
Chiyoda-ku, Tokyo 100, Japan.

3
a Thermopile
c 1.9 kg
d Voltage
e 0 to 20 mV
f 99% response in 45 sec
i 50 m
j $260

4 None

5
a Millivoltmeter
b Potentiometric recorder

7
a Section of Industrial Meteorology, Japan Meteorological Agency, 1-3-4 Ohtemachi, Chiyoda-ku, Tokyo, Japan.
b Lab. of Biology, Dept. of Education, Waseda University, 1-647 Tozuka machi, Shijuku-ku, Tokyo, Japan.

See page xii for Key

1 Pyranometer

2 EKO Instruments Trading Co. Ltd.,*
2-1 Ohtemachi 2-chome,
Chiyoka-ku, Tokyo 100, Japan.

3
a Thermopile
c 2.3 kg
d Voltage
e 0 to 20 mV
f 99% response in 20 sec
i 50 m

4 None

5
a Millivoltmeter
b Potentiometric recorder

See page xii for Key

1 Pyranometer

2 The Eppley Laboratory Inc.,*
12 Sheffield Avenue,
Newport, R.I., U.S.A.

3
a Thermopile
d Voltage
e 7.5 mV per cal cm^{-2} min^{-1}
f 3 – 4 sec
g ca $\pm$ 2%

5
b Potentiometric recorder – 30 millivolts fs

7
a U.S. Weather Bureau (Mr Wright).
b National Research Council,
Toronto, Ontario, Canada. (Dr Latimer).
c Department of Public Health,
Durham, N.C., U.S.A. (Mr. Flowers).

See page xii for Key

1 Precision spectral pyranometer

2 The Eppley Laboratory, Inc.,*
12 Sheffield Avenue,
Newport, R.I., U.S.A.

3
a Thermopile
b Circular receiver 1cm diam.
d Voltage
e 5 mV per cal cm^{-2} min^{-1}
f 1 sec
g ± 1% between –20°and +40° C
h see g; temperature compensation available for –70°to + 50°C
j $990

5
b Potentiometric recorder – 10 millivolts fs

6 Interchangeable filter glass domes for spectral measurements, available at $150 each

7
a U.S. Weather Bureau.
b National Research Council, Toronto, Ontario, Canada.
c Department of Public Health, Durham, N.C., U.S.A.

See page xii for Key

1 Normal incidence pyrheliometer

2 The Eppley Laboratory, Inc.,*
12 Sheffield Avenue,
Newport, R.I., U.S.A.

3
a Thermopile
d Voltage
e 4 - 7 mV per cal cm^{-2} min^{-1}
f 1 sec
j $850

5
b Potentiometric recorder - 30 millivolts

6 Fitted with set of 3 filters for spectral measurements

7
a U.S. Weather Bureau (Mr. Wright).
b National Research Council, Toronto, Ontario, Canada. (Dr. Latimer).
c Department of Public Health, Durham, N.C., U.S.A. (Mr. Flowers).

See page xii for Key

1 Eppley Angstrom pyrheliometer

2 The Eppley Laboratory, Inc.,*
12 Sheffield Avenue,
Newport, R.I., U.S.A.

3
a Electrical compensation
g ± 0.5%

4 Special control unit

5
a Supplied in control unit

6 Standard instrument for calibrating other types of radiometer

7
a U.S. Weather Bureau (Mr. Wright).
b National Research Council, Toronto, Ontario, Canada. (Dr. Latimer).
c Department of Public Health, Durham, N.C., U.S.A. (Mr. Flowers).

See page xii for Key

1 Bimetallic actinograph (No. 58dc.)

2 R. Fuess,
D–1 Berlin 41,
Dunther Str. 8,
Postfach 350, Germany (BRD).

3

a Expansion of black bimetallic element
b 180 mm length x 200 mm depth x 400 mm height
c 6.2 kg
d Chart record
e 0 to 2 cal cm^{-2} min^{-1}
f Slow
g $\pm$ 0.05 cal cm^{-2} min^{-1}
i None
j £178

4 220 V for small vibrator

See page xii for Key

1 Pyranometer

2 Herbert A. Groisse & Co.,*
Rear 1 Gordon Grove, Malvern 3144,
Australia.

3
a Vacuum deposited thermopile (Trickett)
d Voltage
e 0–15 mV·
f 5.6 sec
g + 1%
h - 40°C – + 60°C
i To suit recorder
j $A 165

6 Cosine error < 1% to 70°
Azimuth error < 0.5%
Stability ± 1% in 4 years
(See Norris and Trickett, 1968, p. 245)

7
a University of New South Wales, Australia.
b C.S.I.R.O. Mech. Engineering, Australia.
c C.S.I.R.O. Irrigation Research, Australia.

See page xii for Key

1 Pyranometer

2 Ishikawa Sangyo Co., Ltd.,
4-6-13 Shinkawa, Mitaka City,
Tokyo, Japan.

3
- a Thermopile
- b Diameter of dome 80 mm
- c 1.1 kg
- d Voltage
- e 0 to 10 mV
- f 98% response in 30 sec
- g ± 1.5% of full scale
- h −20° to 40°C
- i 20 to 30 m
- j $250

5
- a Meter to 10 mV
- b Potentiometric recorder 10 mV

7
- a Div. of Observatory, Japan Meteorological Agency, Ohtemachi, Chiyoda-ku, Tokyo, Japan.

See page xii for Key

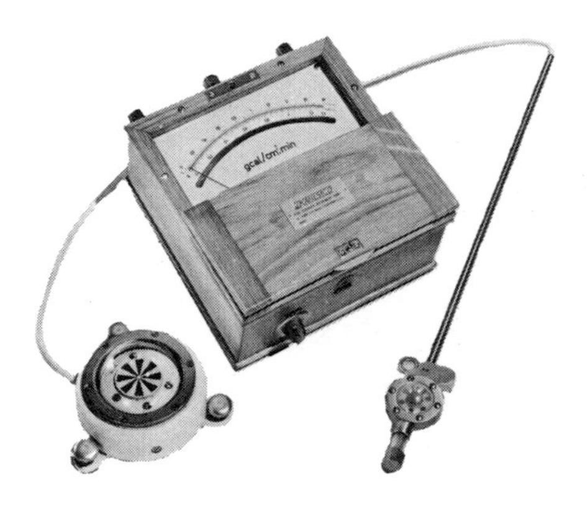

1 Pyranometer

2 Kahl Scientific Instrument Corp.,
P.O. Box 1166, El Cajon,
California 922022, U.S.A.

3
a Thermopile (Dirmhirn)
b 17 cm diam. x 10 cm high
c 0.5 kg
d Voltage
e 1.5 mV per cal cm^{-2} min^{-1}
f 30 sec
j $365

5
a Calibrated moving-coil meter available – see photograph

6 Pyranometer is on left of photograph.
Instrument on right is net radiometer (see p. 108)

See page xii for Key

1 Pyranometer

2 Kipp & Zonen Limited,*
Delft,
Netherlands.

3
a Thermopile (Moll)
b 30 cm diam.; 11 cm tall.
c 3.7 kg in mount, 0.8 kg thermopile housing only.
d Voltage
e 8mV per cal cm^{-2} min^{-1}
f 90% of response in 2.8 sec
g ± 1%
h Temperature coefficient 0.15% per °C
Instrument calibrated at 20 °C
i Maximum length of cable
j £100

6 The instrument can be used with an integrator for totalising solar energy over a given period of time (see p. 96).

See page xii for Key

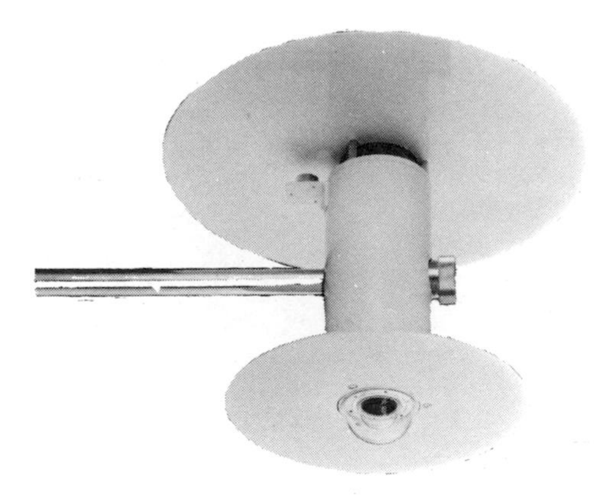

1 Reflectometer

2 Kipp & Zonen Limited,*
Delft,
Netherlands.

3

a Thermopile (Moll)
b Housing 7.5 cm diam.; screens 20 and 30cm diam.
c 6 kg
d Voltage
e 8 mv per cal cm^{-2} min^{-1}
f 10 sec
g ± 1%
j £200

See page xii for Key

1 Pyranometer

2 Lintronic Limited,*
54-58 Bartholomew Close,
London, E.C.1, U.K.

3

a Thermopile
b 6.5 cm diam.; 6 cm high
c 4 oz
d 0.25 mV per mW cm^{-2}
e 0 to 200 mW cm^{-2}
f 30 sec
g Cosine error ± 2%, 0°to 65° incidence;
azimuth error ± 1%
h Fully sealed for under-water working (30 m)
i 100 m
j £25

4 Millivolt recorder or integrator

See page xii for Key

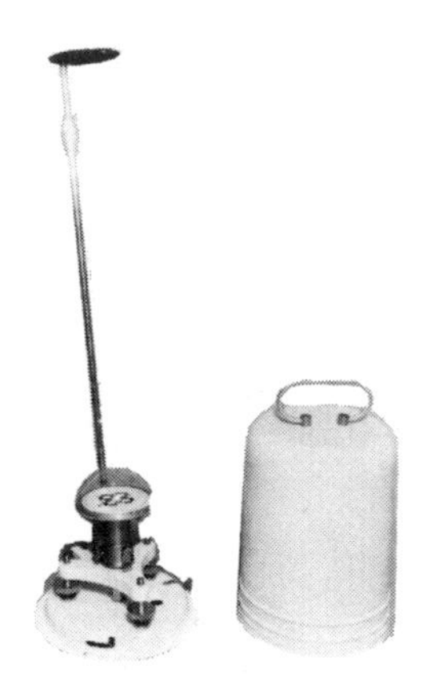

1 Pyranometer

2 Mashpriborintorg,
Smolenskaya–Sennaya Square, 32/34,
Moscow G-200,
U.S.S.R.

3
a Thermopile
b 18 x 18 x 21 cm
c 2 kg
d Voltage
e 7 to 11 mV per cal cm^{-2} min^{-1}
f < 40 sec

5
a millivoltmeter
b millivolt potentiometer

6 The solarimeter can be mounted facing horizontally upwards or downwards and is provided with a shading disc for measurement of diffuse radiation.

See page xii for Key

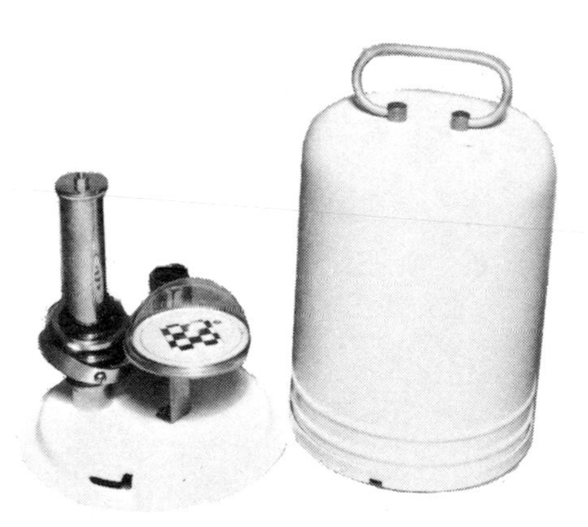

Reflectometer (AP 3 x 3)

Mashpriborintorg,
Smolenskaya–Sennaya Square, 32/34,
Moscow G-200,
U.S.S.R.

Thermopile
18 cm diam. x 23 cm
3 kg
Voltage
7 to 11 mV per cal cm^{-2} min^{-1}
$<$ 40 sec

millivoltmeter

Supplied with gimbal mount

See page xii for Key

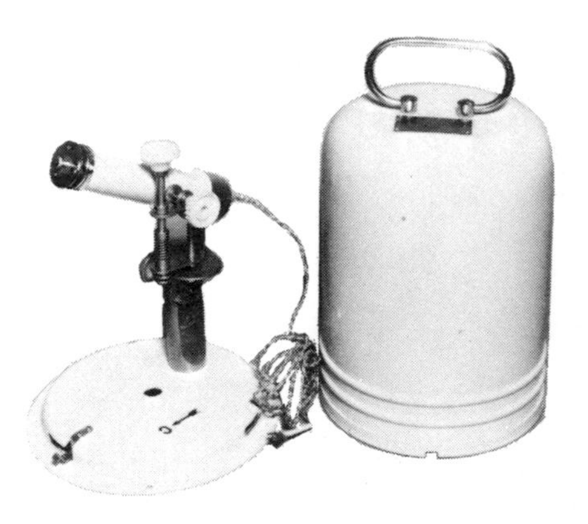

1 Pyrheliometer (AT 50)

2 Mashpriborintorg,
Smolenskaya–Sennaya Square, 32/34,
Moscow G-200,
U.S.S.R.

3

a Thermopile
b 18 cm diam. x 23 cm
c 2 kg
d Voltage
e 5.5 to 7.5 mV per cal cm^{-2} min^{-1}
f < 25 sec

See page xii for Key

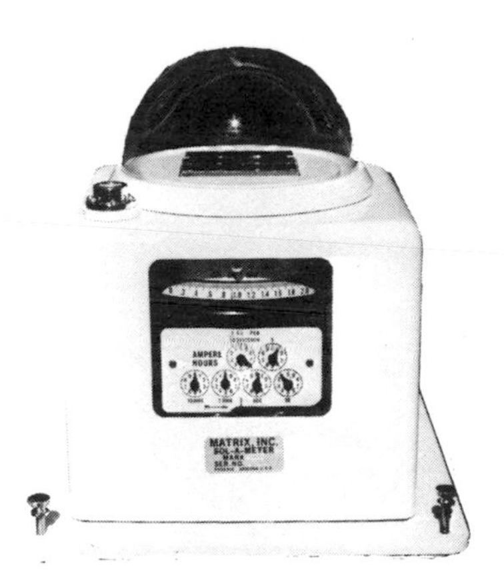

1 Integrating pyranometer (Mark 14)

2 Matrix, Inc.,*
9051 N. 7th Avenue,
Phoenix, Arizona 85021, U.S.A.

3
a Photovoltaic (silicon cell)
b 9 x 11 x 11 in
c 9½ lb
d Current
e 0 to 500 mA or 0 – 2 cal cm^{-2} min^{-1}
Integrator 0 –99,999 Ah calibrated in cal cm^{-2}
f 1 millisecond for 100%
g $\pm$ 5% meter 0.01 cal cm^{-2} min^{-1} ;
integrator 0.01 Ah
h All conditions
j $495

5
a Milliameter
b Milliamp recorder

7
a United States Dept. of the Interior,
Bureau of Reclamation,
Ephrata, Washington, U.S.A.
b Dr. D. Goodwell,
Room 187,
Faculty of Forestry,
The University of British Columbia,
Vancouver 8, B.C., Canada.
c United States Dept. of the Interior,
Bureau of Reclamation,
Regional Office, Region No. 7,
Bldg. 20, Denver Federal Center,
Denver, Colo, 80225, U.S.A.

See page xii for Key

1 Net pyranometer (Mark 18)

2 Matrix, Inc.,*
9051 N. 7th Avenue,
Phoenix, Arizona 85021, U.S.A.

3

a Photovoltaic (silicon cell)
b 18 in diam. 16 in tall
c Approx 20 lb
d Current
e Meter 0 to 500 mA calibrated 0 to 2 cal cm^{-2} min^{-1} integrator 0 – 99,999 Ah calibrated to give total in cal cm^{-2}
f 1 millisecond for 100%
g $\pm$ 5% (meter 2 milliamps, integrator 0.01 Ah)
h All conditions
j $895

6 Integrator incorporated gives time-integral of net solar radiation.

7

a E. A. Hiler, Associate Professor,
Texas A. & M. University,
Agricultural Engineering Department,
College Station, Texas, U.S.A.

b George Peal,
Soil Conservation Service,
P.O. Box 340,
Casper, Wyoming 82601, U.S.A.

c Hollis Shull,
USDA – ARS –SWC ,
Scotts Bluff Station,
Mitchell, Nebraska, U.S.A.

See page xii for Key

Pyranometer (Mark 1-G)

Matrix, Inc.,*
9051 N. 7th Avenue,
Phoenix,Arizona 85021, U.S.A.

Photovoltaic (silicon cell)
5 in diam. x 2 in tall
Approx. 1 lb
Voltage
Standard 0-10 mV; 0 to 2 cal $cm^{-2}min^{-1}$
Approx. 1 millisecond for 100% response
± 5%
Temperature compensated 4°to 60° C
25 ft; or longer if calibrated with cable
$185

0 – 10 mV, impedance, 0.1 ohm
0 – 100 mV, impedance, 1.0 ohm
0 – 10 mV or 0 – 100 mV, impedance same as meter

Filtered units available

7

a Gerald F. Gifford,
Range Science Dept.,
Utah State University,
Logan, Utah 84321, U.S.A.

b James S. Englund, Mech. Engr.,
Engineering Research Division,
Washington State University,
Pullman, Washington, U.S.A.

c Robert W. Owen,
Michael G. Cruse,
Bureau Commercial Fisheries,
P.O. Box 271,
La Jolla, California, U.S.A.

See page xii for Key

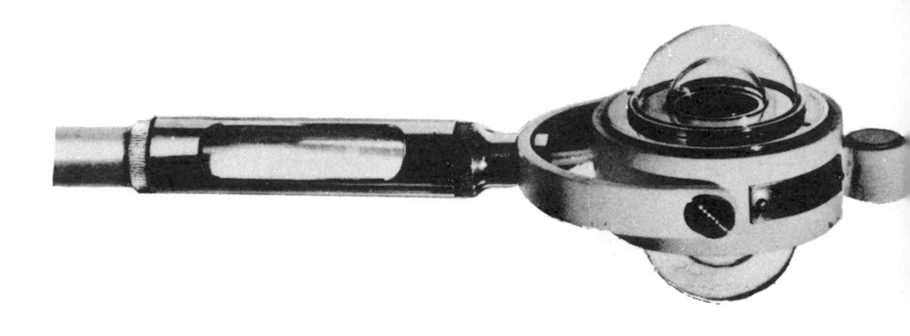

1 Pyranometer

2 Middleton & Co. Pty. Ltd.,
8 Eastern Road,
South Melbourne, Australia.

3
a Thermopile
b 4½ x 4½ x 3¼ in body with 10 in diam. shade disc
c 2 lb
d Voltage
e 0.164 mV per mW cm^{-2}

6 A pair of solarimeters can be supplied mounted back to back for measurement of reflection coefficient; handle includes gimbal mount for levelling (see photograph)

See page xii for Ke

Pyranometer

Nakano-Seisakusho Ltd.,*
Higashimito 1525,
Ueno City, Mie Prefecture, Japan.

Thermopile (Noshi-Denski)
Outside diam. 40 mm
1.5 kg
Voltage
0 – 10 mV
–10°to 50°C
30 m

Millivoltmeter 0-15 mV
Potentiometric recorder 0 – 15 mV

Resistance 25 ohms

Dr. Uemura kenji,
Div. of Meteorology,
National Institute of Agricultural Sciences,
Nishigahara 2-1, Kita-ku,
Tokyo, Japan.

See page xii for Key

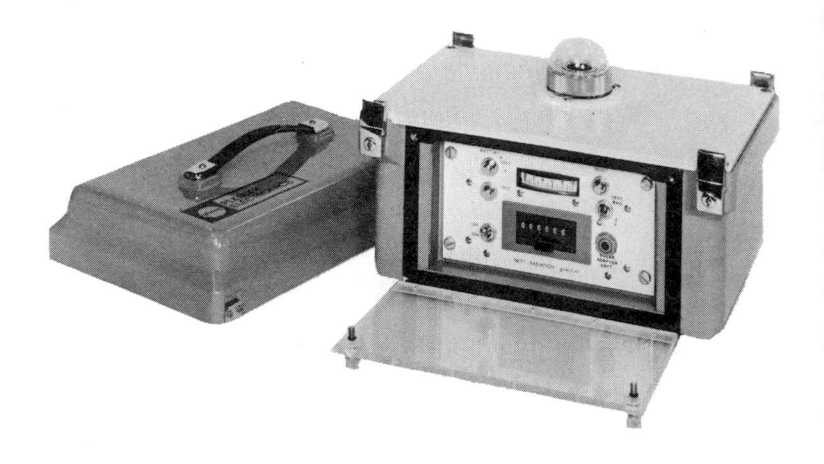

1 Pyranometer integrator

2 Plessey Electronics Ltd.,
Marine Systems Division,
Ilford, Essex, U.K.

3
b 20 x 30 x 20 cm
c 7 kg
e 0 to 1.5 cal cm^{-2} min^{-1}
d Digital counter
e 0 to 1.5 cal cm^{-2} min^{-1}
g $\pm$ 1% fs counter, $\pm$ 2½% fs meter

4 internal battery power supply $\pm$ 12V

5
a Instrument includes Kipp & Zonen pyranometer
(p. 85) mounted in roof of case

See page xii for K

Integrating pyranometer

Rauchfuss Instruments & Staff Pty Ltd.,*
11 Florence Street,
Burwood,
Victoria 3125, Australia.

Silicon solar cell with integrator
Cylindrical, 7 in diam., 10 in high
12 lb
Digital counter
From 4 mW cm^{-2}
$\pm$ 5%; direct reading resolution ca 10mAh
recording resolution 0.2 Ah per contact

reading (indicating only) $ A257
recording, with contact device, $ A337

reading, no power required; recording, power to be supplied by recorder.

reading: built-in meter
recording: strip-chart or digital tape recorders available

6 Portable, no maintenance required other than cleaning the hemisphere occasionally and renewing the silica gel. Useful for collecting data on seasonal and annual totals of incident radiation.

7

Department of Agriculture,
Perth, Western Australia.

b Department of Territories, Papua & New Guinea.

c Commonwealth Bureau of Meteorology, Melbourne, Australia.

See page xii for Key

1 Star pyranometer

2 Philipp Schenk,
A-1212 Wien,
Postfach 3,
Austria.

3
a Thermopile (Dirmhirn)
c 1 kg
d Voltage

5
a Multivoltmeter
b Multivolt potentiometer

6 Intended for measurement of reflection coefficient. The same firm also manufactures the pyranometer shown on p. 99

See page xii for Key

Pyranometer

Philipp Schenk,
A-1212 Wien,
Postfach 3,
Austria.

Thermopile
Voltage

Millivoltmeter
Multivolt potentiometer

This instrument measures the radiation from above and below and on four vertical surfaces

See page xii for Key

1 Solarimeter (S.R.I.3)

2 Solar Radiation Instruments,*
21 Rose St.,
Altona, Victoria, Australia 3018.
(P.O. Box 90 Altona)

3
a Thermopile
b Sensor 1 in diam. shield 8 in diam.
c 5 lb
d Voltage
e 0.35 mV per mW cm^{-2}
f 99% response in 1½ min
g 2½% accuracy
j $A210

5
a Millivoltmeter 0 – 50 mV
b Potentiometric recorder 0 – 50 mV

6 Actinometer or Pyheliometer attachment available for direct observation of solar beam or for checking calibration of other instruments. Pre-calibrated sensors available, matched to S.R.I. standard. Instruments calibrated by C.S.I.R.O. Div. of Met. Physics

7
a N.Z. Department of Civil Aviation
b C.S.I.R.O., Western Australian, Laboratories, Floreat Park, Western Australia.
c Cintru (Physics International), Logue Ave., Mountain View, California, U.S.A.

See page xii for Key

1 Tube pyranometer

2 Swissteco Pty. Limited,
Instrument Division,*
26 Miami Street, Hawthorne East,
Victoria, Australia 3123.

3
a Thermopile
b Diam. 16 mm, length 0.5 m approx.
c 1 lb
d Voltage
f 98% response in 25 sec
g Linear to $\pm$ 1%
h All conditions
i Standard 1 m, any length on request
$A315

5 The protective glass cylinder is easily to exchange; Filters can be mounted between the glass cylinder and the sensor. To prevent condensation the instruments are provided with air inlets and outlets for temporary or permanent flushing with dry air. (See Szeicz, Monteith and dos Santos, 1964, p. 244)

7
a C.S.I.R.O., Div. of Land Research, Canberra, A.C.T., Australia.
Dr. Rose, Dr. Kalma, Mr. Swan.
b The Melbourne University, Botany Dept., Melbourne, Victoria, Australia.
Mr. Friend.

See page xii for Key

1 Pyranometer

2 Yellow Springs Instrument Co.,*
Yellow Springs,
Ohio 45387, U.S.A.

3
a Silicon cell
b Probe 3 x 3½ in; meter 8 x 3 x 5 in; integrator 10 x 11 x 5 in
c Meter 3 lb; integrator 7 lb
d Voltage
e 0 to 2 cal cm^{-2} min^{-1}
f 10 μsec
g 2% cosine response from 0° to 84°
1% from 0° to 50°C; 2% from −25° to 0° and 50° to 50°C
h See g
i 25 ft

4 Power supply 95-125 V, 50-60Hz required for integrator

5
a Meter reads to 2 cal cm^{-2} min^{-1} with resolution of 0.01 cal cm^{-2} min^{-1}
b Integrator reads to 0.1 cal cm^{-2} and responds to 0.005 cal cm^{-2} min^{-1}; incorporates two counters with daily switching and heater thermostat

See page xii for Key

4
Net Radiometers

The **net radiometer** (or net exchange radiometer) is an instrument for measuring the difference between the total radiant flux incident on a surface and the flux reflected and emitted by the surface. The main component of all radiometers is a flat black plate. The temperature difference between the two sides of the plate, measured with a thermopile, is proportional to the net flux of energy received by the plate. The plate must be **ventilated** at a constant rate (e.g. the Gier and Dunkle radiometer) or covered with a **polythene shield** to prevent ventilation (e.g. the Funk radiometer). The performance of both types of instrument is affected by rain and by the accumulation of dust. Standard shielded radiometers have a diameter of about 6 cm but miniature types with diameters down to 1 cm are available for measuring the distribution of net radiation over small animals or leaves.

Tube radiometers are also available for measuring the vertical distribution of flux in crop or forest canopies. They are less accurate than flat plate radiometers and should be used for relative rather than absolute measurements.

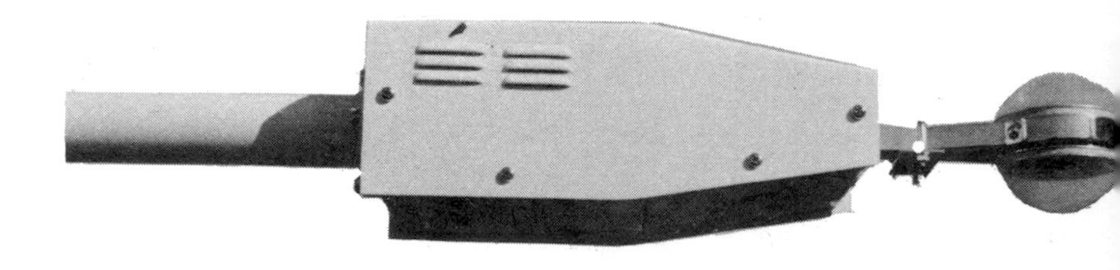

1 Ishikawa net radiometer

2 Ishikawa Sangyo Co. Ltd.,
4-6-13 Shinkawa, Mitaka City,
Tokyo, Japan.

3

a Thermopile
b Diameter of sensor 60 mm
c 10 kg
d Voltage
e 0 to 20 mV
f 45 sec
g ± 3%
h −20° to 45°C
i 30 m
j $350 (RP type), $540 (RLS type)

4 Air pump for circulating dry air into sensor.

5

a mV meter
b Potentiometric recorder – 10 to 40 mV

See page xii for Key

1 Net radiometer (M10)

2 Mashpriborintorg,
Smolenksaya-Sennaya Square, 32/34,
Moscow G-200,
U.S.S.R.

3
a Thermopile
b 20 x 10 x 4 cm
c 0.5 kg
d Voltage
e 7 to 8 mV per cal cm^{-2} min^{-1}
f 12 sec

5
a Millivoltmeter
b Millivolt potentiometer

6 The instrument is unshielded and unventilated. A correction factor for wind speed is provided with the calibration certificate.

See page xii for Key

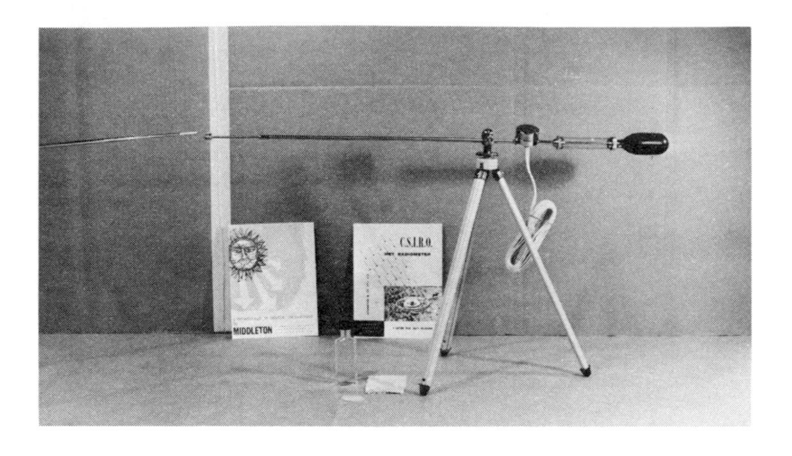

1 Net radiometer (CN 6)

2 Middleton & Co. Pty. Ltd.,
8 Eastern Road, South Melbourne,
Australia.

3
a Miniature thermopile (Funk)
b 12 mm diameter
d Voltage
e 3 mV per cal cm^{-2} min^{-1}
j \$A300

6 Internal resistance 40 ohms.
(see Funk, 1962, p. 245)

See page xii for Key

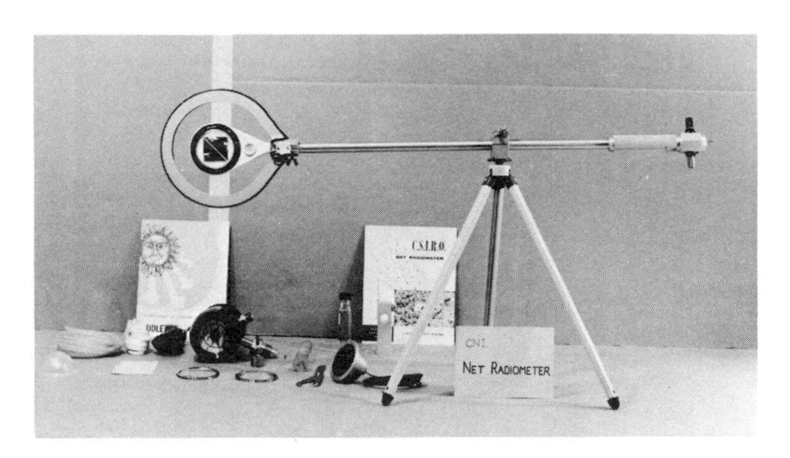

1 Net radiometer (CN 1)

2 Middleton & Co. Pty. Ltd.,
8 Eastern Road, South Melbourne.
Australia.

3
a Thermopile (Funk)
b 16 x 18 cm
d Voltage
e 30 mV per cal cm^{-2} min^{-1}
f 90 sec
g $A370

6 Heating ring available to prevent
condensation on external surface
of polythene domes. Interior flushed
with nitrogen. Internal resistance 80 ohms.
(See Funk, 1962, p. 245)

See page xii for Key

1 Net radiometer

2 Philipp Schenk,
A-1212 Wien,
Postfach 3,
Austria.

3
a Thermopile (Sauberer-Dirmhirn)
b Domes 28 mm diam.; 40 cm long
c 0.5 kg
d Voltage
e 6mV per cal cm^{-2} min^{-1}
f 40 to 60 sec
j £130

6 Perspex cylinder holding silica gel
attached for drying.
Also available from Kahl Scientific
Instrument Corporation (see p. 84)

See page xii for Key

1 Net radiometer

2 Science Associates Inc.,
Box 230 Nassau St.,
Princeton, N.J. 08540,
U.S.A.

3
a Thermopile (Gier & Dunkle)
c 15 lb
d Voltage
e 2.5 mV per cal cm^{-2} min^{-1}
f 12 sec for 95% response
g ± 1% from –20° to 160°F
h see g
j $850

5
a Millivoltimeter
b Recorder with 7 in scale available; chart speed 1 in per hr $775

6 Includes temperature compensation circuit. Instrument can be adapted for measuring total hemispherical radiation.

See page xii for Key

1 Net radiometer

2 Science Associates, Inc.,
230 Nassau St.,
Box 230,
Princeton, N.J. 08540, U.S.A.

3
a Thermopile (Fritschen)
b 6 cm diam.
c 4 oz
d Voltage
e 3.5 mV per cal cm^{-2} min^{-1}
f 12 sec
h Temperature compensated
from 10 to 50°C
j $260

5
a Millivoltmeter
b Millivolt recorder

6 Pressurization system available, $40
(see Fritschen, 1965, p. 246)

See page xii for Key

Miniature net radiometer (S.R.I.7)

Solar Radiation Instruments,*
P.O. Box 90 Altona, Victoria,
Australia 3018.

Thermopile (Funk)
0.5 in diameter
5 oz
Voltage
0.15 mV per mW cm^{-2}
99% response in 30 sec
$\pm$ 5%
100 m approx
$A 170

Millivoltmeter 25 mV fs
Potentiometric recorder 25 mV fs

Gas supply G.S.1. available for portable operation otherwise G.S.2. gas supply (electric) or G.S.3 industrial gas cylinder. Adaptor attachment for uni-directional measurements. Calibrated by C.S.I.R.O. Div. of Met. Physics, Aspendale, Vic.
(see Funk, 1962, p. 245)

7

a Professor J.L. Monteith,
University of Nottingham,
School of Agriculture,
Sutton Bonington, nr. Loughborough,
Leicestershire.

b UN Food and Agriculture Organisation
(F.A.O.), University of Guelph,
Ontario, Canada.

See page xii for Key

1 Net radiometer (S.R.I. 4)

2 Solar Radiation Instruments,*
P.O. Box 90, Altona,
Victoria, Australia 3018.

3
- a Thermopile (Funk)
- b 2 in diam.
- c 14 oz
- d Voltage
- e 0.85 mV per mW cm^{-2}, 190 ohms; or 0.45 mV per mW cm^{-2}, 45 ohms
- f 99% response in 1 min
- g ± 2½%
- i 100 m approx
- j $A220

5
- a Millivoltmeter 50 or 100 mV fs
- b Potentiometric recorder 50 or 100 mV fs

6 Three types of gas supply available G.S.1. Portable, G.S.2. electric, G.S.3 industrial gas cylinder. Heating ring or ventilating rings available for dew dispersion, solarimeter attachment and uni-directional adaptor attachment.

Calibrated by C.S.I.R.O. Met. Phys
Aspendale, Victoria, Australia.
(see Funk, 1962, p. 245)

7
- a Professor J.L. Monteith, University of Nottingham, School of Agriculture, Sutton Borington, nr. Loughborough, Leicestershire.
- b UN Food & Agriculture Organisatic (F.A.O.), University of Guelph, Ontario, Canada.
- c Centre du National Recherche, Paris, France.

See page xii for Key

1 Miniature net radiometer

2 Swissteco Pty. Limited,
Instrument Division,*
26 Miami Street, Hawthorn East,
Victoria, Australia 3123.

3

a Thermopile (Funk)
b Sensor diam. 10 mm, body diam. 14 mm
c Approx 0.5 lb
d Voltage
f 98% response in 25 sec
g Linear to ± 1%, balanced to ± 3%,
h Accuracy of calibration ± 2½%
i Standard 1 m any length on request
j $A280

5 Polythene hemispheres easy to exchange. Inside ventilation or closed circuit with silica gel container to keep instrument dry. Sensor designed as an interchangeable cartridge, easy to exchange for servicing.
(see Funk 1962, p. 245)

See page xii for Key

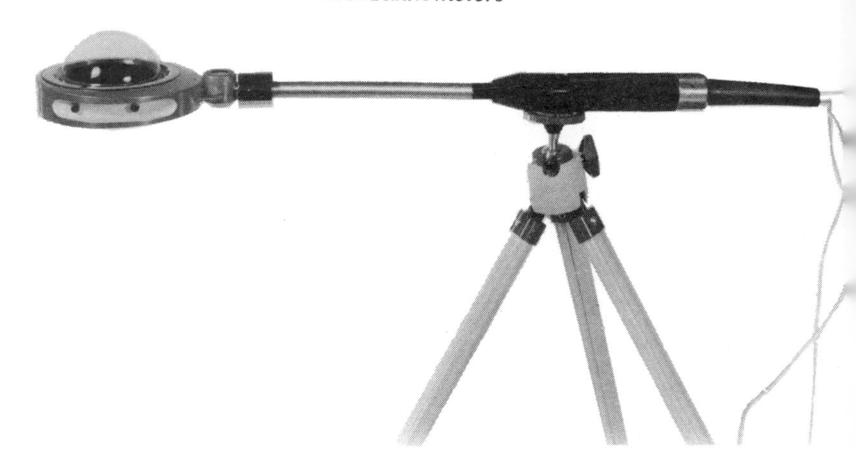

1 Net radiometer (S1)

2 Swissteco Pty. Limited,
Instrument Division,*
26 Miami Street, Hawthorn East,
Victoria, Australia 3123.

3

a Thermopile
b Sensor 50 mm diam., body 95 mm diam.
d Voltage
e 0 – 50 mV
f 98% response in 25 sec
g Linear to ± 1%; calibration accurate to ± 2½%
h All conditions
i Standard 1 m, on request any length
j $A 325

6 Polythene hemispheres easy to exchange. Inside or closed circuit ventilation to keep instrument dry. Outside air curtain to prevent dew, and to keep hemispheres free from rain drops and snow flakes. No heating ring necessary.

Adaptor for uni-directional reading (bla
body to cover one side of instrument)
Also available as Short Wave Balance Me
(SW-1) equipped with two double glass domes. (see Funk, 1962, p. 245)

7

a McGill University, Geography Dept., Montreal, Canada. Dr. T.R. Oke.
b Kansas State University, Evapotranspiration Laboratory, Manhattan, Kansas, U.S.A. Prof. E.T. Kanemasu.
c The University of Wisconsin, College of Agriculture, Madison, Wisconsin, U.S.A. Prof. C.B. Tanner.

See page xii for Key

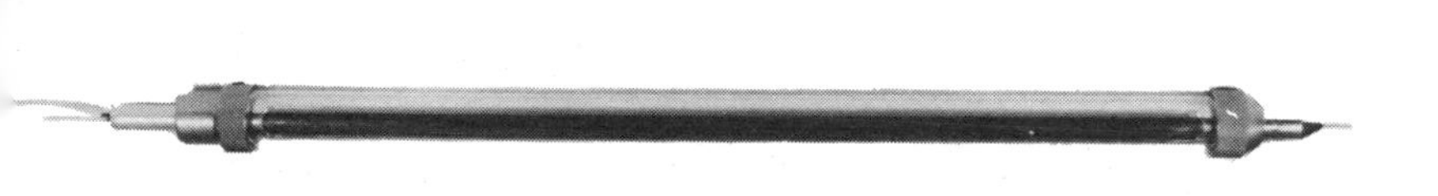

1 Tube net radiometer

2 Swissteco Pty. Limited,
Instrument Division,*
26 Miami Street, Hawthorn East,
Victoria, Australia 3123.

3
a Thermopile
b Diam. 1.6 cm, two models 1 m and 0.5 m long
d Voltage
e 0 – 100 mV
f 98% response in 25 sec
g Linear to ± 1%
h All conditions
i Standard 1 m, on request any length
j $A 297 (1 m long)
$A 233 (0.5 m long)

6 The tubular net radiometers can be dismantled for easy servicing. The protective wind shields (Polythene tubes) are easily exchanged. To prevent condensation build up, the instruments are provided with air inlets and outlets for temporary or permanent flushing with dry air

These instruments are also supplied as tubular Net Pyranometers (Short Wave Balance Meters) with interchangeable double glass cylinders. $A 330 (1m long), $A 265 (0.5 m long)

7
a Kansas State University,
Evapotranspiration Laboratory,
Manhattan, Kansas, U.S.A.
Prof. E.T. Kanemasu.
b C.S.I.R.O., Div. of Land Research,
Canberra, A.C.T., Australia.
Dr. Rose, Dr. Kalma.
c State of Israel, Dept. of Agriculture,
The Volcani Institute of Agriculture,
Israel. Prof. G. Stanhill.

See page xii for Key

1 Net radiometer

2 Teledyne Geotech,
3401 Shiloh Road,
Garland, Texas 75050, U.S.A.

3
a Thermopile (Gier and Dunkle)
c 33 lb
d Voltage
e −2 to +3 mV
g Calibration ± 2%, temperature compensation
h ± 1%
j $935

4 Power supply 115V 50 to 60 Hz
or 12 V DC

5
a Millivoltmeter
b Potentiometric recorder

See page xii for Key

1 Net radiometer

2 C.W. Thornthwaite Associates,*
Elmer, New Jersey, 08318, U.S.A.

3
a Thermopile
b 5 cm diam. x 21.5 cm long including mounting stem
c 85 g
d Voltage
e 0.25 or 3.5 mV per cal cm^{-2} min^{-1}
f 3.3 sec
i 25ft with Model 603 recorder; 500 ft with high impedance recorder.
j $175 (0.25 mV), $335 (3.5mV)

5
a Millivoltmeter (see photograph)
b Calibrated battery-operated recorder available, $900 including radiometer, purge kit and 30 m lead

See page xii for Key

1 Miniature net radiometer

2 Toa gijutsu center,*
4-11-5 Kotobashi,
Sumida-ku, Tokyo, Japan.

3
a Thermopile
b 3 x 3 x 1 cm sensor
c 0.9 kg
d Voltage
e 10 mV per cal cm^{-2} min^{-1}
f 10 sec
h 0° to 60° C
i 30 m
j $115

4 Dry battery

5
a Millivoltmeter –3 to +7 mV
b Potentiometric recorder –3 to +7 mV

7
a Prof. Mihara Yoshiaki, Dept. of Horticultu
Chiba University, Matsudo City,
Chiba pref., Japan.

See page xii for Key

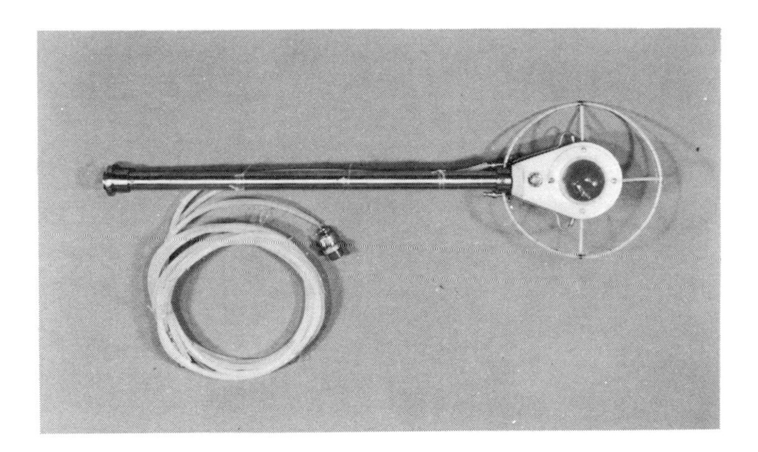

Net radiometer

Toa gijutsu center,*
4-11-5 Kotobashi,
Sumida-ku, Tokyo, Japan.

Thermopile
Sensor 3 cm diam.
1 kg
Voltage
0 to 30 mV per cal cm^{-2} min^{-1}
10 sec
0.3 mV
0°to 60°C
$210

Flushing with dry air is needed

Millivoltmeter 50 mV f s
Potentiometer

Sensor should be set in precisely horizontal position.

7
a Dr. Ozawa yukio, National center for Natural disaster prevention, Higashi-Ginza, Minato-ku, Tokyo, Japan.

See page xii for Key

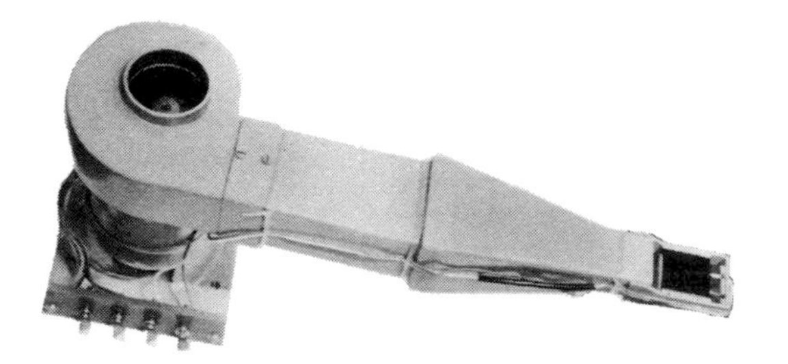

1 Net radiometer

2 Toa gijutsu center,*
4-11-5 Kotobashi,
Sumida-ku, Tokyo, Japan.

3
a Thermopile
b Body 15 x 15 x 30 cm ;
sensor 3 x 3 cm
c 2 kg
d Voltage
e 8 mV
f 10 sec
h 0°to 60° C
i 30m
j $170

4 Power supply (100V, 5A)

7
a Prof. Mihara Yoshiaki, Dept. of Horticulture,
Chiba University, Matsudo City, Chiba pref.,
Japan.

See page xii for Key

5
Other Radiation Instruments

Most of the radiation instruments included in this section are designed to measure either surface radiative temperatures or the spectral distribution of solar radiation. Instruments for measuring long-wave radiation (3 to 100 μm) are called **pyrgeometers** by the World Meteorological Organisation, and are distinguished from **pyrradiometers** which measure the total flux of short and long-wave radiation on a horizontal surface. Neither of these terms is *commonly* used. The older and simpler term **pyrometer** describes an instrument for measuring the radiation emitted by a surface and hence its radiative temperature. The total flux received on a horizontal surface e.g. from the sun and sky, is usually measured by covering one side of a ventilated or shielded radiometer with a metal plate.

Spectral radiometers incorporate either a set of coloured glass filters for measuring the amount of radiant energy in relatively broad wavebands (e.g. 100 nm), or wedge interference filters providing a continuous spectrum (effective bandwith about 10 nm).

Other more specialised instruments include a photon meter for measuring the quantum content of solar radiation and an ultra-violet illuminometer. Conventional light meters which measure radiant energy weighted by the response of the human eye are not included as commercial instruments but several relevant papers will be found among the references.

1 Radiometer for surface temperature measurement (PRT 10L)

2 Barnes Engineering Company,
30 Commerce Road,
Stamford, Connecticut 06902, U.S.A.

3

a Measurement of infra-red radiation between 7 and 20 μm

b 25 cm long x 21 cm high x 9 cm wide (hand held with pistol grip)

c 0.9 kg

d Voltage

e Absolute scale –10°to +60° C, differential scale ± 5° C

f 90% response in 2 sec

g 0.2° C at 20°C

h as e

j $1200

6 Field of view 35° or 7°
This instrument is completely self-containe
and portable. The manufacturer makes a range of radiometers operating between –4
and 1100° C
(see Gates, 1968, p. 247)

See page xii for Key

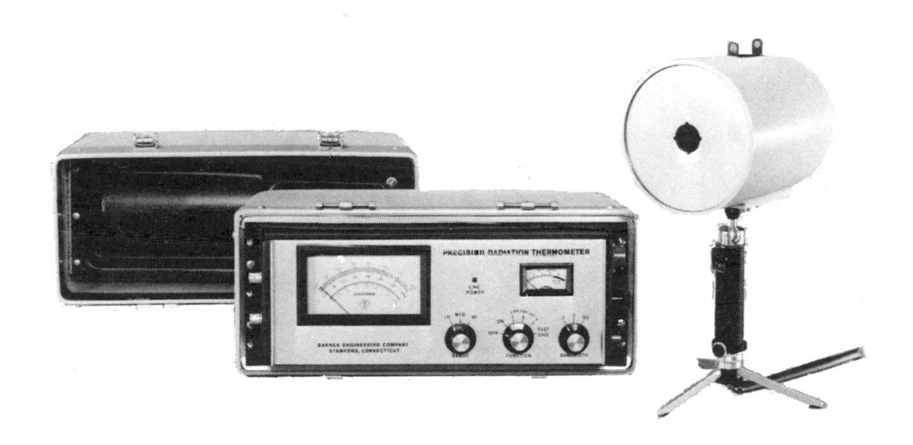

1 Precision Radiation Thermometer (PRT–5)

2 Barnes Engineering Company,*
30 Commerce Road,
Stamford, Connecticut 06904, U.S.A.

3
a Measurement of infra-red radiation
b In carrying case 16 x 16 x 16 in
c With case 35 lb
d Meter display with output jacks for remote recording
e Temperature range –20° C to +75° C other ranges available
f 5 milliseconds
g 0.5° C accuracy with better than 0.1° C resolution
h –20° C to +40° C
i 100 ft
j $7,645

4 Self contained batteries, rechargeable from 115 or 230 volts, 50-400 Hz supply 20 watts

5
a Meter display with output jacks for remote recording

6 Portable, easy to operate; for ground as well as airborne temperature measurement

7
a Dr. D. Lorenz,
Deutscher Wetterdienst,
Meteorologisches Observatorium,
8126 Hohenspeissenberg,
Albin-Schwaiger-Weg 10, Germany (BRD).

See page xii for Key

1 Light integrator (SRF–458)

2 Iio Denki Ltd.,*
Yoyogi 2-27-18, Shibuya-ku,
Tokyo, Japan.

3
a Silicon cell with interference filter
b sensor 40 x 82 cm; integrator 15 x 21 x 30 cm
c Sensor 250 g; integrator 7 kg
d Digital counter
g $\pm$5%, $\pm$ 0.01 cal cm^{-2}
h –10°to 50° C; 0 to 80% rh
j $695

4 Power supply 100V, 50/60 Hz

5
a On a day with rain or snow, it is hoped
that the measurement stops

See page xii for Key

1 Recording spectrophotometer

2 Iio Denki Ltd.,*
Yoyogi 2-27-18, Shibuya-ku,
Tokyo, Japan.

3

Monochromator
Sensor 27 x 36 x 18 cm; amplifier 27 x 53 x 23 cm; recorder 27 x 53 x 30 cm
Sensor 15 kg; amplifier 25 kg; recorder 20 kg
0.3 to 0.4, 0.4 to 0.7, 0.7 to 0.9μm
$\pm$ 5% of indication at 0.55μm
0°to 50°C; 0 to 80% rh
10 m
$6000

See page xii for Key

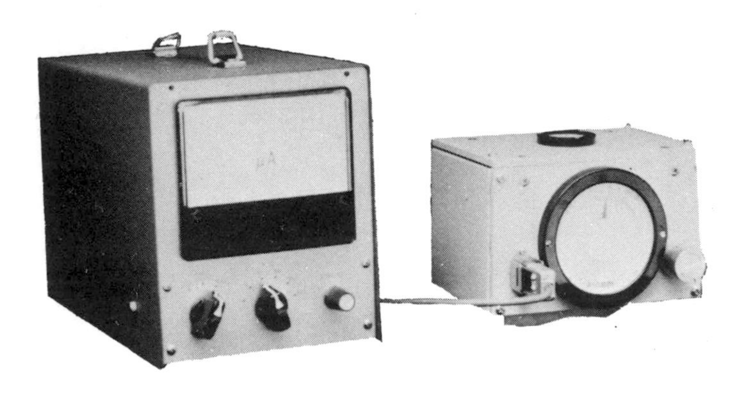

1 Portable spectrophotometer

2 Iio Denki Ltd.,*
Yoyogi 2-27-18, Shibuya-ku,
Tokyo, Japan.

3
a Silicon detector with interference filters
b Sensor 11 x 18 x 12 cm; indicator 14 x 30 x 25 cm
c Sensor 7 kg, indicator 4 kg
d Current
e 0.4 to 0.7μm
energy flux from 20 to 2 x 10^3 μW cm^{-2} fs
g $\pm$ 10%
h −10° to +50°C; 0 to 80% rh
i 2m
j $1,250

4 100 V, 50/60 Hz

5
a ammeter
b potentiometric recorder 0 to 10 mV

6 This equipment is portable and applicabl for measurement of radiation regime wit crop canopies. It is hoped that the senso and indicator keep in a laboratory under rain conditions.

See page xii for Key

1 Spectroradiometer

2 E.G. and G.,
Electronic Products Division,
160 Brookline Avenue,
Boston, Mass. 02215, U.S.A.

3
a Monochromator and range of photodetectors
b Voltage
c 0.2 to 1.2 μm
d Calibration accuracy $\pm$ 8% from 0.35 to 1.2 μm

4 Power supply 110 V 50/60 Hz or internal battery

5
a Integral meter
b Output for recorder provided

6 Calibrations provided for all grating – detector combinations. Fibre optic probes and telescope systems available.

See page xii for Key

1 Ultra-violet radiometer

2 The Eppley Laboratory Inc.,
12 Sheffield Avenue,
Newport, R.I. 02840, U.S.A.

3

a Selenium photocell

d Voltage

e 0.2 mV per mcal cm^{-2} min^{-1}

f Milliseconds

g Linear to $\pm$ 2%; cosine corrected to $\pm$ 2.5% for 0° to 80°

h –40°to +40° C

5

b Potentiometric recorder – 30 millivolts fs

6 1000 ohms impedance. Interference filter incorporated limits response to 295 to 385 nm waveband

7

a U.S. Weather Bureau (Mr. Wright).

b National Research Council, Toronto, Canada (Dr. Latimer).

c Department of Public Health, Durham, N.C., U.S.A. (Mr. Flowers).

See page xii for Key

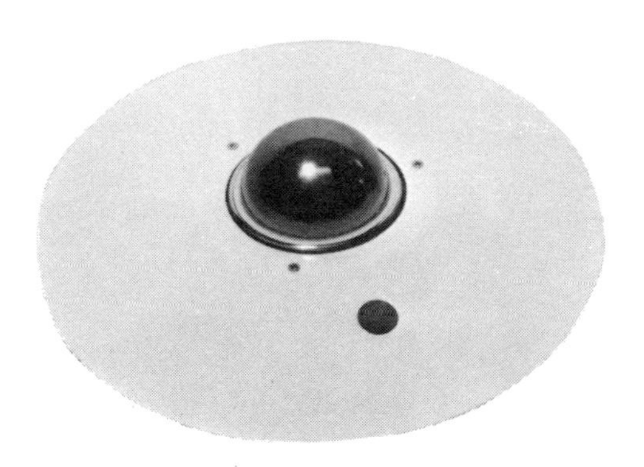

Pyrgeometer

Eppley Laboratory Inc.,
12 Sheffield Avenue,
Newport, R.I. 02840, U.S.A.

Thermopile
Voltage
5 mV per cal cm^{-2} min^{-1}
2 sec
Cosine error less than 5%; linearity $\pm$ 1%
$1000

The instrument is similar to the precision spectral pyranometer but is fitted with a KRS - 5 dome which transmits radiation from 4 to 50μm.
The thermopile circuit includes temperature compensation to give output proportional to incident long-wave flux.

See page xii for Key

1 Spectral radiometer (QSM 2400)

2 Incentive Research and Development,*
Box 11074,
S-161-11 Bromma,
Sweden.

3
a Photocell and filter
b Measuring unit 10 x 10 x 36 cm
Control unit 17 x 11 x 8 cm
d Voltage
e 400 to 700 nm
f less than 0.1 sec
g better than 1% for band width of 13 nm
h 0°to 50° C; below 0° C with heater. Can be used below water to 80 m depth
i 150 m
j £950 without chart recorder

4 4 of 6.75 V, 2 of 2.7 V mercury cells

5
a not incorporated
b millivolt potentiometer

6 Completely portable

7
a Dept. of Botany,
Lund University,
Sweden.

b Dept. of Agriculture and Fisheries,
Marine Laboratory,
Aberdeen.

c Dept. of Nature Resources,
Canada.

See page xii for Ke

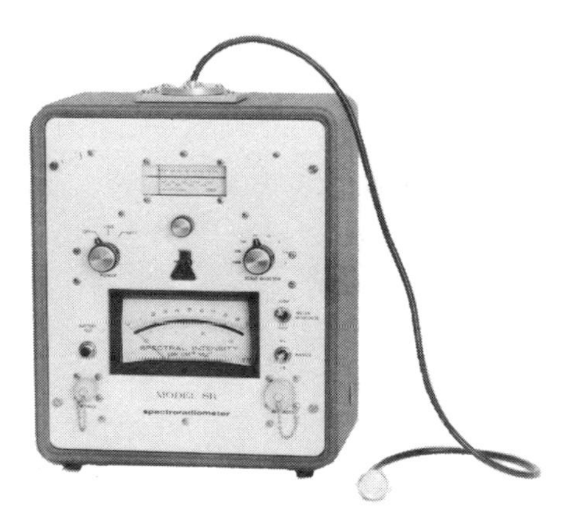

Spectroradiometer

Instrumentation Specialities Company Incorporated,*
Lincoln, Nebraska 68507, U.S.A.

Photodiode with wedge interference filter
12 x 10 x 7 in
12.5 lb
Voltage
380-750 nm or 380-1350 nm
2 sec
7% to 10%, band width 15 nm
380 - 750 nm £1820 duty paid
380 - 1350 nm £1935 duty paid

8 ranges from 0.3 to 10^3 μW cm^{-2} nm^{-1} fs
Potentiometric recorder, 10 mV

Portable for field use
Available 110 V, 50 Hz
220 V, 50 Hz

c Glass fibre optic remote probe enables measurement in confined spaces. Available in 3ft, 6ft and 12ft lengths.
d Accessories: Programmed scanner and spectral lamp calibrator.
e Teflon diffusing screen gives true cosine response.

7 Supplied on request

See page xii for Key

1 Light meter (23 CM 120)

2 Kahl Scientific Instrument Corporation,
P.O. Box 1166,
El Cajon,
California 92022, U.S.A.

3
a 3 photodiodes with set of 3 filters
b Probe 40 mm diam. x 15 mm;
case 31 x 17 x 9 cm
c 1.1 kg complete
d Voltage
e 0.4 to 0.5 μm ('blue') 0.6 to 0.7 μm
('red') 0.7 to 0.8 μm ('far red')
i 4 ft

5
a Meter graduated 0.5 to 10 $\mu W\ cm^{-2}\ mm^{-1}$
with x 10 and x 100 switched ranges

6 Fully transistorized and powered by 9 volt
battery.

See page xii for Key

1 Universal radiation meter

2 Kahl Scientific Instrument Corporation,
P.O. Box 1166,
El Cajon,
California 92022, U.S.A.

3
a Thermopile (Moll-Gorczynski)
d Voltage
e 7mV per cal cm^{-2} min^{-1}
i 50 m

5
a −100 to +400μA incorporated

6 The instrument is supplied as a self contained field kit with adaptors for measuring total, direct or diffuse solar radiation, atmospheric radiation net radiation, etc. RG2, OG1 and WG7 filters are provided.

See page xii for Key

1 Pyrheliometer, pyrgeometer (Linke-Feussner)

2 Kipp & Zonen Ltd.,*
Delft, Netherlands.

3
a Compensated thermopile
b Overall height approx. 30 cm
c 6 kg
d voltage
e 11 mV per cal cm^{-2} min^{-1} ; 1° C temperature difference generates 0.7μV
f 8 sec
j £480

5
a Millivoltmeter

6 Instrument measures direct solar radiation or long-wave radiation in cone with half angle of 5°. Provided with set of filters. Resistance of thermopile 65 ohms.

See page xii for Key

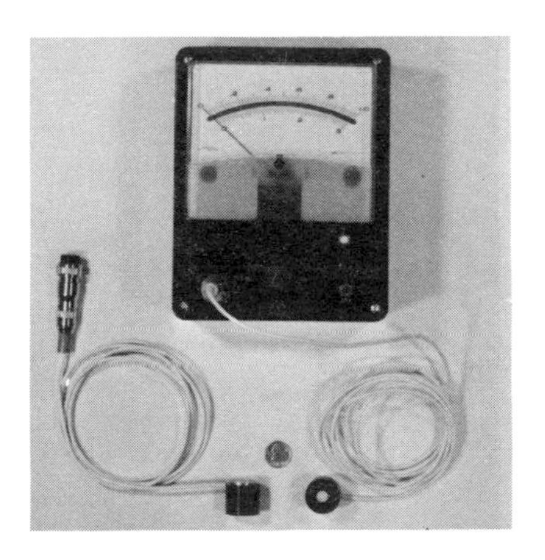

1 Photon meter

2 Lambda,
2933 North 36th,
Lincoln,
Nebraska 68504, U.S.A.

3
a Silicon cell with filter
b Sensor 1.6 x 1.6 cm; meter 14 x 16 x 9 cm
c 2 lb (meter)
d Current
e Sensor 50μA per μEinstein cm^{-2} sec^{-1}
meter 10^{-3} to 1μEinstein cm^{-2} sec^{-1} fs
g 1% fs
j Sensor $140; meter $375

5
a Meter incorporated, integrator available

6 Response is proportional to wavelength and therefore to number of quanta per unit wavelength. Other sensors are available for measuring illumination ($60) or total solar radiation ($40) (see Norman, Tanner and Thurtell, 1969, p. 248)

See page xii for Key

1 Radiation thermometer

2 A. Levermore,
40-44 Broadway,
London, S.W. 19., U.K.

3
a Thermopile
d Voltage
e 15° to 40° C
g Sensitivity ± 0.25° C on meter
± 0.1° C on recorder; reproducibility
0.5° C
h 18° to 32° C
j £990

4 Standard mercury batteries (500 hours
operation)

5
a incorporated
b output for recorder

6 Accepts radiation from area of 4 mm diam.
at 25 mm distance. Intended for skin
temperature measurement.

See page xii for Key

1 Light meter

2 Metra Blansko,*
Blansko, Hybesova 53,
Czechoslovakia.

3
a Photocell
b 15 x 9 x 5 cm
c 0.45 kg
d Voltage
e 10 to 10^5 lux
h –10°to 40°C, 45 to 80% rh
i 15 m

5
a Meter incorporated

7
a Tesla Holesovice,
Ing. Behal,
Brno,
Tatranska 12.
b Vedecko vyzkumny uhelnyustav,
Ing. Siska,
Ostrava – Radvanice.

See page xii for Key

1 Pyrometer (Ardonox)

2 A. G. Siemens,
D-8250 Erlangen 2,
Postfach 325,
Germany (BRD).

3
a Thermopile
b 8 cm diam. x 18.5 cm long
c 1.5 kg
d Voltage
e 0°to 60°C; –0.085 to +0.232 mV
f 2.1 sec
g $<$ 2% fs
h 0°to 60°C
j £145

4 Stabilized power supply £21

5
a Microvoltmeter
b Microvolt recorder

6 Minimum measuring circle is 9 cm diam.
at distance of 10 cm

See page xii for Key

1 Photometer

2 T.F.D.L., Dr. S.L. Mansholtlaan 12, Wageningen, Netherlands.*

3

a Selenium cell with opal-glass for true cosine response and neutral filter for linearity
b Head 67 mm diam., 30 mm high and handle 22 mm diam., 150 mm long
c 0.4 kg
d Current
e Short-wave irradiance up to 1200 W/m^2
f 10 msec
g Accuracy ± 5% (–10°to +50° C) resolution ± 0.5%
h Temperature –20°to +60° C humidity max. 70% for long periods, for a hermetically sealed version 100% admissible.
i About 100 m
j About £60 (without meter)

4 None

5 micro-ampere meter or recorder with full scale ranges of (preferably) 1 μA to 1 mA and a potential drop of about 10 mV (not more than 35 mV).

6 Portable, used for light measurements in the open field, in the crop and in growth chambers

7 Agricultural University, Wageningen, Netherlands.

See page xii for Key

1 Integrating photometer

2 T.F.D.L., Dr. S.L. Mansholtlaan 12, Wageningen, Netherlands.*

3

a Integration of photo-current from a vacuum photo-tube (Rank Electronics type VB 39)

b 195 x 68 x 88 mm

c 1.4 kg

d Current

e Total short wave irradiation 0.5 to 1200 W/m^2

f 10^{-4} s

g Accuracy ± 5%, resolution dependent on the integrator

h Intended for field use; temperatures −10° to +50° C, humidity has no effect

i Several meters of cable with high isolation resistance

j £220 (including electronic integrator and batteries)

4 Integrator and batteries

5 Integrated irradiation presented on an electro-mechanical counter

6 Self-contained, portable for use in field and plant canopies. Provided with diffuser for cosine response and filter for uniform sensitivity to radiant energy in waveband 400-700 nm

7 Agricultural University, Wageningen, Netherlands. (Several departments).

See page xii for Key

1 Photon meter

2 T.F.D.L., Dr. S.L. Mansholtlaan 12, Wageningen, Netherlands.*

3
a Differential measurement from two groups of silicon cells measuring visible and infrared and infrared only
constant quantum response 400–700 nm
b Two versions: head 67 mm diam., 30 mm high, handle 22 mm diam., 150 mm long; or 23 mm high and wide, 900 mm long, handle 200 mm
c Short type: 0.4 kg
long type: 1.3 kg
d Voltage
e Up to about 1.5. 10^{21} photons m^{-2} sec^{-1} in the 400–700 nm range
f 0.1 msec
g Accuracy ± 5% (−10°to +50°C), resolution ± 2%; cosine corrected
h Temperature −20°to +60°C
no humidity restriction
i About 100 m
j Short type about £70 (without mV meter)
long type about £580 (without mV meter)

4 None

5 mV meter or recorder with suitable ranges (highest range needed is 25 mV). Input resistance at least 10 k ohms

6 Both types are portable; used for measurement of photosynthetically active radiation.
Short type : general use
Long type : use in crops

7 Agricultural University, Wageningen, Netherlands.

See page xii for Key

1 Radiation thermometer

2 T.F.D.L., Dr. S.L. Mansholtlaan 12, Wageningen, Netherlands.*

3
- a Thermopile
- b Length 185 mm (including handle), 90 mm high, 42 mm wide
- c 1.1 kg
- d Voltage
- e Application usually in the range of environmental temperatures, about −20° to +70° C
- f 8 sec
- g Accuracy ± 0.2° C in the range 0°–50° C resolution ± 0.1° C
- h Same as e., operation not affected by humidity
- i No essential limitation
- j About £70 (matching microvoltmeter about £190)

4 Known temperature (e.g. ice-point) for thermocouple measurement of the temperature of the reference radiator. The microvoltmeter is battery operated.

5
- a Microvoltmeter with resolution of 0.5 μV Full scale ranges: 25, 100, 250 μV, 1 and 2.5 mV
- b Recorder with same resolution or 2.5 mV full scale type for use with the recorder output of the microvoltmeter. Both instruments should have a high input resistance (min. 10^5 ohm).

6 Portable, used for temperature measurements of leaves, soil, trees, crops, open water etc.
See Stoutjesdijk (1966), p. 247

7 Agricultural University, Wageningen, Netherlands.

See page xii for Key

1 Radiometer

2 Yellow Springs Instrument Co.,*
Yellow Springs,
Ohio 45387, U.S.A.

3
a Compensated thermistor bolometer
b 10 x 6 x 8 in
c 8 lb
d Bridge voltage
e Seven ranges available from 0 to 0.25 mW cm^{-2} – 0 to 250 mW cm^{-2}
f 99% response 20 sec
g Accuracy 5% fs; resolution 0.5% fs
h Compensated within the range 18°to 40° C humidity to 95% rh
i 10 m
j $525 including readout

4 Mains operated

5
a Millivoltmeter incorporated
b 0–10 mV potentiometer with input impedance of at least 10 k ohms

6 Sensitive from 0.28 to 2.6 μm

7 Available on request

See page xii for Key

6

Heat Flow Meters

A heat flow meter is a device for measuring the flow of heat through material such as soil or across a surface such as the skin of an animal. It consists of a thermopile giving an output proportional to the temperature difference across a plate. Ideally, the plate should have the same thermal conductivity as the surrounding medium but this is rarely possible in practice. Alternatively, the plate should be as thin as possible to avoid distorting the temperature field in its neighbourhood. Soil heat flux plates must be thoroughly waterproofed and are usually buried within a few centimetres of the surface. For medical or animal work, thin flexible plates are cemented to the skin with rubber solution.

1 Heat flow meter

2 Keithley Instruments, S.A.,
14 Avenue Villardin, Ch-1009 Pully,
Switzerland.

3

a Thermopile
b 17 x 39 x 0.4 mm
d Voltage
e Max. 320 W/m^2
g $\pm$10%
h Calibration curve supplied for $-300°$ to $+400$ °F
j Sensor $89
Sensor and DC amplifier unit $525

5

a DC amplifier is calibrated in 7 switched ranges from 0.32 to 320 W/m^2 fs
b Recorder output $\pm$ 1V from 5a

6 Thermal resistance of plate is 0.2° C per W/m^2. Flexible enough to be used on curved surface.

See page xii for Key

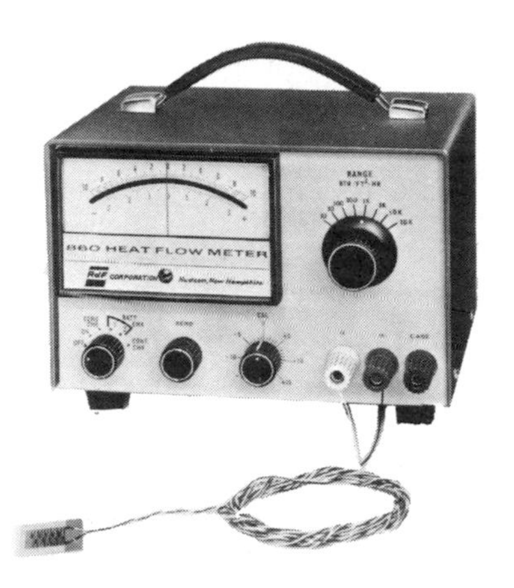

1 Heat flow meters

2 RdF Corporation,
23 Elm Avenue, Hudson,
New Hampshire 03051, U.S.A.

3
a Thermopile
b From 0.3 x 0.5 in to 1 x 1 in
From 0.003 to 0.012 in thick
d Voltage
e From 0.007 to 2.7 μV per W/m^2
f **20 to 40 msec**
h −300° to +500° F
j $45 to $185

5
a Microvoltmeter
b Microvolt potentiometer

6 Ten thermocouple configurations available. All sensors made from flexible polymide film which can be surgically implanted to study heat flow in living tissue.

See page xii for Key

1 Heat flow meter

2 Science Associates Inc.,
Box 230 Nassau St.,
Princeton, N.J. 08540,
U.S.A.

3
a Thermopile
b Sensor 6 cm x 2.8 mm
Recorder 18 x 23 x 25 cm
d Voltage
e 11 μV per W/m^2
i 6 ft standard
j Sensor $96; recorder $695

5
a Microvoltmeter
b Recorder will operate on rechargeable batteries for 200 h. Centre zero reading $\pm$ 5mV adjustable to + 10 mV.

6 Thermal conductivity of sensor similar to conductivity of dry soil. Electrical resistance 200 ohms.

See page xii for Key

1 Heat flux meter (S.R.I.9)

2 Solar Radiation Instruments,*
21-21A Rose St., Altona, Melbourne,
Victoria, Australia.

3
a Thermopile
b 2 in diam. x ⅛ in thick
c 1 oz
d Voltage
e 30 μV per W/m^2
g $\pm$ 5%
i 50 m approx.
j $50

5
a Microvoltmeter
b Microvolt potentiometer

6 Calibrated by Div. of Met. Physics,
Aspendale, Victoria.

7
a University of Aberdeen, Scotland.
b University of East Anglia, U.K.
c Maquarie University, N.S.W., Australia.

See page xii for Key

1 Heat flow meters

2 T.F.D.L., Dr. S.L. Mansholtlaan 12,
Wageningen, Netherlands.

3
a Thermopile
b 13, 23, 50, 106 mm diam.
d Voltage
e 17 to 220 μV per W/m^2 depending on size
h Two smaller sizes are silicone rubber, max. 400°C; two larger sizes are PVC, max. 70°C

5
a Microvoltmeter
b Microvolt potentiometer

6 Thermal resistance is 4.3×10^{-3} °C per W/m^2 for silicone rubber discs 2.5 mm thick and 1.7×10^{-2} °C per W/m^2 for PVC discs, 3.3 mm thick

See page xii for Key

1 Heat flow meters

2 Thermonetics Corp.,
Thermal Instrumentation Specialists,
P.O. Box 9112, San Diego,
California 92109, U.S.A.

3

a Thermopile
b 12 sizes from ½ x ½ in to 2¼ in x 2½ in
d Voltage
e From 2μV to 3 mV per W/m^2
f 1 sec
g to ± 1%
h −325° F to 2000°F depending on type
j $70 to $310

5

a Microvoltmeter/millvoltmeter
b Potentiometer

See page xii for Key

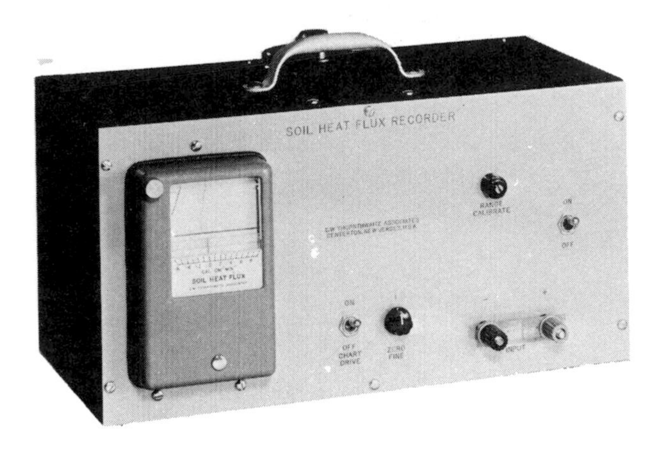

1 Heat flow meter (Model 610)

2 C.W. Thornthwaite Associates,
Elmer, New Jersey 08318, U.S.A.

3
a Thermopile
b 25 mm diam., 2.6 mm thick
d Voltage
e 6μV per W/m^2
i Standard lead 1 m
j $85

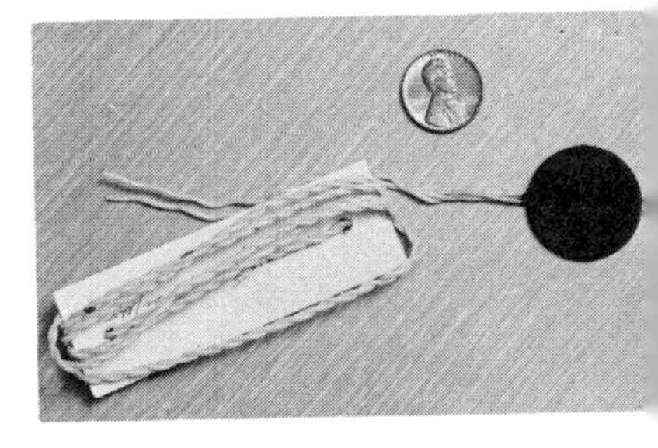

5
a Microvoltmeter
b Calibrated battery operated recorder available at $800 complete with sensor and 30 m cable

See page xii for Key

Heat flow disc

Ernest Turner,
Chiltern Works, High Wycombe,
Bucks, U.K.

Thermopile
12 mm diam., 1.5 mm thick
Voltage
0.7 μV per W/m^2

Microvoltmeter
Microvolt potentiometer

Disc of tellurium-silver alloy with copper
gauze on two sides.

See page xii for Key

7
Anemometers

The two main types of instrument used in micrometeorological studies are **cup anemometers** and **hot-wire anemometers.** The rotation of a cup anemometer can be used to operate an electrical contact directly or to interrupt a beam of light falling on a phototransistor. Several phototransistor instruments incorporate all the associated electronics in the anemometer housing and have stopping speeds in the range 5 to 10 cm/s. Similar mechanical contact models have stopping speeds of 20 to 40 cm/s. In both types, the speed of rotation increases almost linearly with wind speed above the stopping speed.

Hot-wire anemometers must be provided with a source of constant current or constant voltage, or must be kept at a constant temperature. They operate down to windspeeds of a few centimetres per second and can be used in enclosed or restricted spaces where cup anemometers will not rotate. Their main disadvantages are fragility and an output which is not linearly related to windspeed. **Heated thermocouple** anemometers are more robust and are available commercially. Heated resistance coil and heated thermistor anemometers are relatively easy to make in the laboratory but there are very few models on the market.

Pressure tube and **vane anemometers** are seldom used in micrometeorological studies because their response depends on orientation with respect to wind direction. They are intended for flow measurements in wind tunnels and ducts where the direction of flow is uniform and well defined.

1 Cup anemometer

2 E.H. Bernfeld Ltd.,*
282 Kingsland Road,
London E.8., U.K.

3
a Generation of current
b Cup circle 10 cm diam., height 21 cm
c 0.4 kg
d Current
e 0 to 10; 0 to 50 m/sec
g $\pm$ 2%, stopping speed 25 cm/sec
h Applicable in tropical climates

7
a Hanseatische Werkstatten, Friedrichs & Co.,
2 Hamburg, 61,
Olderloerstr 97/99, Germany (BRD).
b Wilhelm Lambrecht,
34 Gottingen, Friedlander Weg, 65/67,
Germany (BRD).
c Friebe Luftfahrtbedarf,
69 Heidelberg,
Postfach 530, Germany (BRD).

See page xii for Key

Cup anemometer (Sheppard type)

C.F. Cassella & Co. Ltd.,*
Regent House,
Brittania Walk,
London, N.1., U.K.

Rotating cups
30 x 23 x 10 cm
2 kg
12v square pulse
0.1 to 15 m/sec
–40°to +55° C
300m
£55

12 volt DC Battery

Digital counter and circuit available

University of Nottingham,
Department of Agricultural Sciences,
School of Agriculture,
Sutton Bonington,
Loughborough, Leicestershire, U.K.

b Rothamsted Experimental Station,
Harpenden,
Hertfordshire, U.K.

c Wye College of Agriculture,
Ashford,
Kent,
U.K.

See page xii for Key

1 Cup anemometer (Robinson type)

2 Iio Denki Ltd.,
Yoyogi 2-27-28, Shibaya-ku,
Tokyo, Japan.

3
a Rotation of cups
b 7.5 x 21 x 19 cm
c 2.3 kg
d One contact closure per 100 m run
g 2.5%, 2m/s
h −10°to 50° C, 0 to 80% rh
j $390

4 Power supply 100 V, 50/60 Hz

See page xii for Key

Hot wire or hot film anemometer

Disa Elektronik A/S,
Hervlev, Denmark.

Cooling of heated element
Minimum probe size 5 μm diam., 1 mm long
Bridge voltage
to 150 m/sec
up to 150°C
50 m
From £129 for basic constant temperature unit; most probes in range £5 to £15

Voltmeter range 0 – 10 volts
Potentiometric recorder

System includes wide range of probes, linearizers, power supplies, digital voltmeters, calibration equipment, etc. Facilities for micrometeorology include analogue signal for autocorrelation function and probe for measuring x y z components of velocity.

See page xii for Key

1 Thermocouple anemometer indicator/
recorder

2 Hastings – Raydist Inc.,
Hampton, Virginia 23361, U.S.A.

3

a Temperature difference of heated and unheated thermocouples

b Probe 12 in long x 2 in diam.; recorder 4 x 4 x 9 in

c 7 lb

d Voltage

e 0 to 15 m/sec

i 100 ft

j $370

4 115V, 60Hz

5

b Chopper-bar type, records every 5 sec

6 Also available with meter only and 12 ft cable – $210

See page xii for Ke

Thermocouple anemometer

Hastings – Raydist Inc.,
Hampton, Virginia 23361, U.S.A.

Temperature difference of heated and unheated thermocouple
Probe 12 in long x 2 in diam.; meter 14 x 10 x 7 in
8 lb
Voltage
0 to 500 and 500 to 10,000 ft/min
$490

115V, 60Hz

Precision millivoltmeter with 7 in scale incorporated

Also available with directional probe $75

See page xii for Key

1 Thermistor anemometer

2 Light Laboratories,*
10 Ship Street Gardens,
Brighton, NB1 1AJ,
Sussex, U.K.

3

a Temperature difference of heated and unheated thermistor
b Probe 3 x 9 cm; case 25 x 14 x 15 cm
c 2.5 kg
d Voltage
e 0 to 100, 0 to 500, 0 to 1,000 ft min^{-1}
g ± 5% low ranges; resolution 10 ft on low ranges, 50 ft on high ranges
h to 70° C, 80% rh
i 100 ft
j £59

4 9V battery

5

a DC microammeter, 200 mA 350 ohms pivoted instrument

7

a Royal Aircraft Establishment,
Mr. Winkless,
Farnborough, Hants, U.K.

b Royal Air Force,
O/C, Halton, Aylesbury,
Bucks, U.K.

c Babcock & Wilcox Ltd.,
Medical Department (Dr. Ross),
Renfrew,
Scotland.

See page xii for Key

Hot film anemometer

Lintronic Ltd.,*
54/58 Bartholomew Close,
London, E.C.1., U.K.

Cooling of hot platinum foil
Probe dimensions 4 mm x 10 cm long,
2 mm x 2.5 cm long
25 g
Voltage
Linear from 0 to 500 m sec^{-1}
Frequency response flat from 0 to 100 kHz
0.1% of range

110 or 230 V, 50 or 60 Hz

5
1 V, 1 kohm meter
Oscilloscope, etc.; output impedance
is 8 kohms

6 Measures mean flow under turbulent
conditions. microscale turbulence, etc.

7
a S.C. Ling,
Space Science Department,
Catholic University of America,
Washington D.C., U.S.A.
b N.G.T.E.,
Farnborough,
Hants, U.K.

See page xii for Key

1 Cup anemometer (Makino)

2 Makino Oyosokki Co. Let.,*
19-4 Numabukuro 3-chome,
Nakano Oku, Tokyo, Japan.

3
a Rotation of cups
b Cup diam. 5 cm; arm length 6 cm; height 20 cm
c 300 g
d Voltage pulse
e 0.2 to 30 m/sec
g ± 0.2 m/sec
h −10° to 40° C (wind sensor)
−20° to 60° C (counter unit)
i 50 m
j $2000

4 AC power and DC power are applicable for performance of this instrument

5
a Wind indicator 0 to 30 m/sec
b pulse counter (KE-5M)

6 The remote indication is feasible to 200 in distance with extension cable betwee wind sensor and the pulse divider. The digital recording is available by replacin the counter unit with the printer.

7
a Dr. Uchijima Zenbei,
Division of Metereology,
National Institute of Agricultural Scien
Nishigahara 2-1, Kita-ku, Tokyo, Japan
b Prof. Takeda Kyoichi,
Lab. of Agricultural Meteorology,
Faculty of Agriculture,
Kyoshu University,
Hakozaki, Fukuoka, Japan.

See page xii for Key

1 Anemometer (M27, M 27c)

2 Mashpriborintorg,
Smolenskaya-Sennaya Square, 32/34,
Moscow G-200,
U.S.S.R.

3
a Force on hemispheres
b Sensor 23 cm diam. x 28 cm; recorder 67 x 34 x 33 cm
c Recorder 65 kg
d Bridge voltage
e 5 to 90 m sec^{-1}
i 100 m

5
b Oscillograph (M27) and chart recorders (M27c) are available

6 A system of 21 small metal hemispheres exerts torque on a metal tube which is measured with strain gauges

See page xii for Key

1 Anemometer (M 61)

2 Mashpriborintorg,
Smolenskaya-Sennaya Square, 32/34,
Moscow G-200,
U.S.S.R.

3
a Rotation of cups
b 7 x 21 x 29 cm
c 0.6 kg
d reading of dial
e 1.5 to 40 m sec^{-1}
h –40° to 45° C

6 A clockwork mechanism incorporated in the base of the anemometer stops rotation automatically 5 min after the starting mechanism has been operated.

See page xii for Key

1 Anemometer and direction recorder (M 47)

2 Mashpriborintorg,
Smolenskaya-Sennaya Square, 32/34,
Moscow G-200,
U.S.S.R.

3
a Rotation of vane and displacement
b Transmitter 60 x 55 x 29 cm;
receiver 28 x 20 x 14 cm
c Transmitter 4.5 kg; receiver 3.5 kg
d Current
e 1.5 to 50 m sec^{-1} , $0°$ to $360°$

5
a Meters incorporated

6 Rotating vanes are linked to
generator and direction vane to
servo-system

See page xii for Key

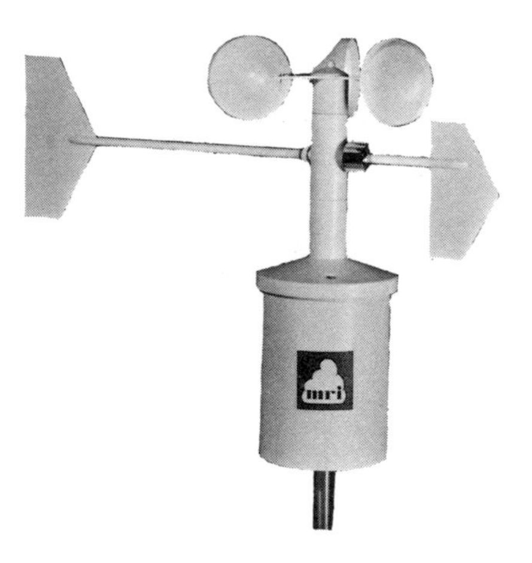

1 Cup anemometer and wind vane (2040)

2 Metereology Research Inc.,
464 W. Woodbury Road,
Altadena,
California, U.S.A.

3
a Rotation of cups
d Recorder chart
g Resolution of wind run 0.1 mile
j $3,425; components available separately

5
b strip chart recorder (Esterline Angus)

6 Wind speed, azimuth, and standard deviation of azimuth angle are recorded simultaneously on chart.

See page xii for Key

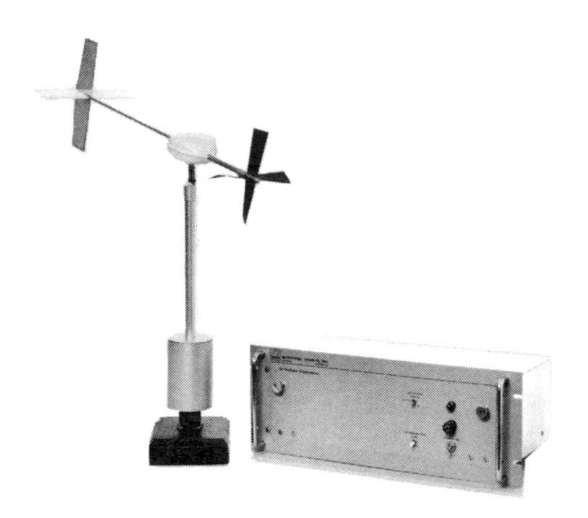

1 Bivane anemometer (1053)

2 Meteorology Research Inc.,
464 W. Woodbury Road,
Altadena,
California, U.S.A.

3

a Rotation of propeller & displacement of vane
b Propeller 6 in diam.; 20 in high
c 1½ lb
d Voltage
e Azimuth 0°-540°, elevation ± 60°; 0.5 to 50 ms^{-1}
f Response distance 1 m
g Linear to ± 1% for direction and speed; resolves azimuth and elevation to 1°
j $2,700

4 Transmitter for processing signals included in 3j

5

b Strip chart recorder

6 Mounting arm and calibration facilities available

See page xii for Key

1 Thermocouple anemometer
and thermometer

2 Nihon Kagaku Kogyo Co. Ltd., *
Tokyo,
Japan.

3
a Cooling of heated thermocouple
b Probe of 11 mm diam.x 20 cm long
Control unit 22 x 29 x 11 cm
c Control unit 4 kg
d Voltage
e 0 to 5 and 5 to 40 m/s; –50° to 150° C
f ca 30 sec
i 5 m standard

4 Battery operated

5
a Meter incorporated

See page xii for Key

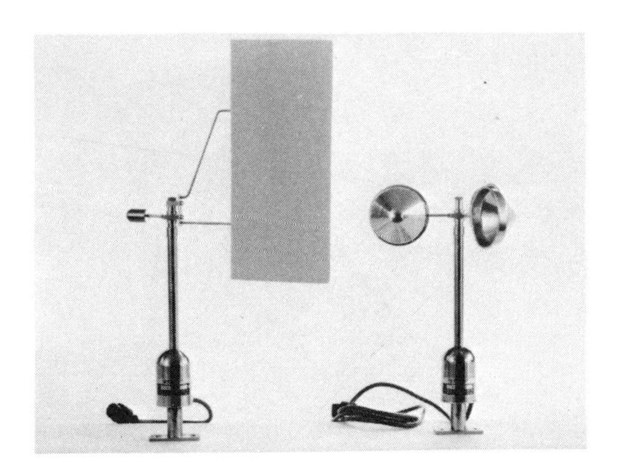

1 Cup anemometer

2 Rauchfuss Instruments & Staff Pty. Ltd.,*
11 Florence Street,
Burwood,
Victoria, Australia 3125.

3
a Rotating cups
b Cup diam. 3¼ in, cup circle 10 in, height 13 in
c Cup 4 g, cup assembly 40 g
d 12V pulse or contact closure
e Stopping speed ca 10 cm/sec
g ± 1%, linear within ± 1½%
h −25°to +60° C, 0 to 95% rh
i equivalent to 50 ohms
j Pulse type $A165 contact type $A149

4 12V DC supply

5
a Digital counter
b Strip chart or digital event recorder available

6 Low torque instrument for use in low wind speeds. Matching direction indicator available – see photograph.
(see Sumner 1968 p. 249)

7
a Division of Meteorological Physics,
C.S.I.R.O.,
Aspendale,
Victoria, Australia.
b Division of Plant Industry,
C.S.I.R.O.,
Deniliquin,
NSW, Australia.
c University of Brisbane,
Queensland,
Australia.

See page xii for Key

1 Cup anemometer

2 Rauchfuss Instruments & Staff Pty. Ltd.,*
11 Florence Street,
Burwood,
Victoria, Australia 3125.

3
a Rotating cups
b Cup diam. 1¼ in outside cup circle 4 in, height 4 in
c Cups 2.8 g, complete 28 g
d One volt pulse or contact closure per revolution
e Pulse type stops at 15 to 20 cm s^{-1}
Contact type stops at 30 to 35 cm s^{-1}
g 1%
h −25° to 60° C; 0 to 95% rh
i Equivalent to 50 ohms
j Pulse type $A145 including counter
Contact type $A147 including counter

4 12V DC supply

5
a Digital counter
b Strip chart or digital event recorder available

6 Available singly or as set of six in box with counters, cables and 12V battery (ca $A700) (see photograph) (see Bradley, 1969, p. 249)

7
a Research School of Biological Sciences, A.N.U.,
Canberra, Australia.
b Forest Research Institute,
Canberra, Australia.
c Scientific & Industrial Research,
Wellington,
New Zealand.

See page xii for Key

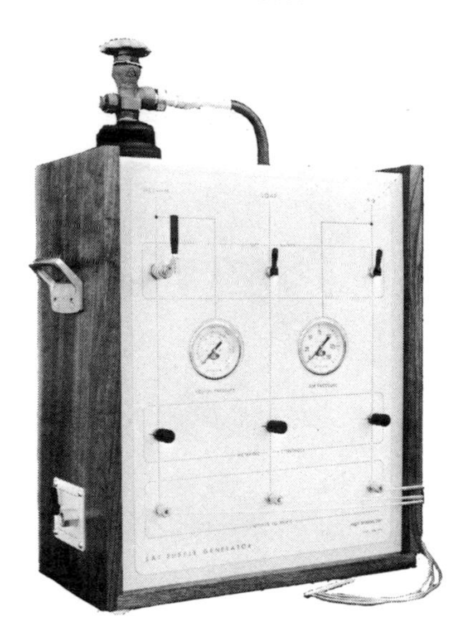

1 Flow visualization system

2 Sage Action Inc.,
P.O. Box 416,
Ithaca, N.Y. 14850, U.S.A.

3
a Movement of soap bubbles
b Bubbles $\frac{1}{16}$ in to $\frac{1}{4}$ in diam.; generator 21 x 18 x 9 in
c 38 lb
d Stream of bubbles

4 Compressed air and helium cylinders

6 Bubbles are generated at a rate between 1 and 20 per sec and can be photographed to determine air velocity and direction

See page xii for Key

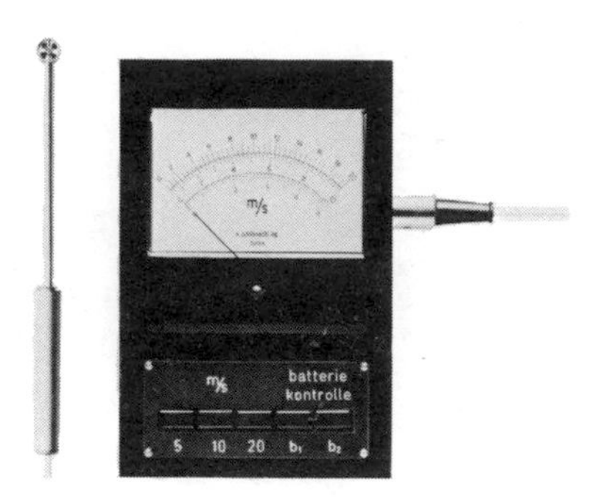

1 Miniature vane anemometer

2 E. Schiltknecht Ingenieur,
8047 Zurich,
Freilagerstrasse 11,
Switzerland.

3
a Rotation of vanes
b Vane diam. 2 cm
d Current proportional to pulse frequency
e 0–2, 0–6 and 0–20 m sec^{-1}
Stopping speed 0.3 m sec^{-1}
g $\pm$ 1.5% fs at 20° C
h –30°to +65° C
i 1.2 m standard, longer if needed

4 9V battery supply

5
a Calibrated meter supplied, 9 cm scale

See page xii for Key

1 Generator anemometer

2 E. Schiltknecht Ingenieur,
8047 Zurich,
Freilagerstrasse 11,
Switzerland.

3
a Rotation of cups
b Overall cup diam. 10 cm, overall length 14 cm
d A.C. voltage and frequency
e 0.6 to 50 m/sec; 4 V and 16Hz at 10 m/sec
h −30°to +65°C

5
a A.C. voltmeter

6 Voltage and frequency are both linear functions of velocity. Interfaces with digital and BCD outputs available.

See page xii for Key

1 Cup anemometer

2 Spembly Ltd.,
Newbury Road,
Andover,
Hampshire, U.K.

3
a Rotation of cups
b Overall height 28 cm, body diam. 6.2 cm
c 340 g
d Voltage from internal ratemeter or frequency
e 0.25 to 15 m/sec gives 0 to 9 V DC
f 0.1 sec for change from 4 to 6 m/sec
g ± 1%
h –40° to +50° C
i At least 1000 m

4 Power supply 220/240 V 50Hz or battery pack available as extra

5
a Voltmeter 0 to 9 V
b Recorder output provided

6 A matching wind direction indicator is available; also main display and share display units for wind speed and direction.
Intended for short and long-term studies of airflow
(see Jones, 1970, p. 250)

See page xii for Key

1 Semi-conductor anemometer

2 T.F.D.L., Dr. S.L. Mansholtlaan 12, Wageningen, Netherlands.*

3

a Hot bulb principle; heated semi-conductor temperature sensor with reference sensor

b 2.5 mm sphere

c 2 g

d Voltage

e 0–1.5 m/s

f 15 sec

g Accuracy $\pm$ 0.2 m/s; resolution better at low velocities

h 0°– 40°C, the sensor must be protected against rain

i Cable length not limited

j £87 for sensor with reference; electronic circuit £140

4 Battery operated

5

a 200μA moving coil meter reading. Conversion of meter reading to wind velocity by means of graph

b Electronic circuit has recorder output for potentiometric 10 mV or galvanometric recorders 200 μA

6 Portable, used for micrometeorological wind velocity measurements, e.g. near the ground, in glass-houses in crops.

7 Agricultural University, Wageningen, Netherlands.

See page xii for Key

1 Thermocouple anemometer

2 H. Tinsley & Co. Ltd.,
Werndee Hall,
South Norwood,
London, S.E. 25, U.K.

3
a Temperature difference of heated and unheated thermocouple
b Meter 31 x 23 x 21 cm
sensor 2.5 cm long x 0.5 mm diam.
c Meter 5.4 kg
d Voltage
e 0 to 1.5 m sec^{-1}

i Standard 120 cm
j £170

5
a Meter incorporated

6 Maintains calibration within close limits for long periods

See page xii for Key

1 Hot wire anemometer

2 Toa gijutsu Centre,*
4-11-5 Kotobashi,
Sumida-ku, Tokyo, Japan.

3

a Cooling of gold-coated tungsten wire
b 50 mm
c 4 kg
d current
e 0 –3 m/sec
f 1 sec
g 2 cm/sec
h 0°to 50°C
i 3 m
j $250

5

a Meter supplied

7

a Prof. Sakagami Jiro, Dept. of Science,
Ochanomizu University, Otsuka,
Bunkyo-ku, Tokyo, Japan.

See page xii for Key

1 Anemometer with thermometer (GGA 235)

2 Wallac Oy, *
Turku 5, Finland.

3
a Hot wire
b 23 x 10 x 6 cm, hand-held meter plus probe
c 1.1 kg
d Bridge voltage
e Temperature −30°to +30°, +20°to +80°, +60° to +180° C
Wind 0.1 to 5, 2 to 30 m/sec
f 50% response in 0.2 sec
g Temperature 1% fs, wind 8% of mid-scale reading
h −30°to + 150° C, 15 to 90% rh
i 3m
j £100

5
a Incorporated
b External recorder available

6 Direct-reading battery operated instrum[ent]
A series of special temperature probes i[s] available, e.g. for surface temperature measurement.

7
a W.R. Curtsinger,
Ohio State University, U.S.A.

b Dept. of Physiology and Environmental Studies,
University of Nottingham,
Sutton Bonington,
Loughborough, Leicestershire, U.K.

See page xii for Key

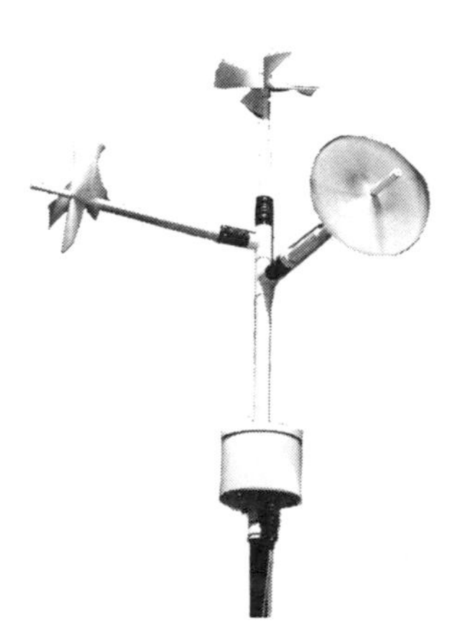

1 3 Dimensional anemometer (27002)

2 R.M. Young Company,*
42 Enterprise Drive,
Ann Arbor, Michigan 48103,
U.S.A.

3
a 3 propellers measure xyz wind components
b Base housing with three sensors mounted at right angles 42 in high
c 8 lb
d Analog voltages – three bipolar signals from DC tachometer generators
e 50–0–50 miles/hour
f Distance constant 3.1 feet (approx. 2 cps)
g ± 1%, cosine conformity 10%, infinite resolution
h –30°C to + 70°C, 0–100% rh
i 1000 feet (10 conductors)
j $696

4 Three channel recorder or data logger

5
a -500 – 0 – 500μA
b Millivolt or 5 mA galvanometer (3 channel)

6 Produces analog voltages (bipolar) directly proportional to wind vectors. Semi-portable; normally requires AC voltage but can be used without line voltage. Sensors are self generating using miniature tachometer generators. Signal polarity reverses when wind direction reverses. Especially suitable for computer data reduction. Very low threshold and fast response for mechanical type anemometer.

7
a Battelle Northwest,
Richland; Washington, U.S.A.
b White Sands Missile Range,
New Mexico, U.S.A.
c Kennedy Space Flight Center,
N.A.S.A.,
Florida, U.S.A.

See page xii for Key

1 Anemometer Bivane (21002)

2 R.M. Young Company,*
42 Enterprise Drive,
Ann Arbor, Michigan 48103,
U.S.A.

3
a Rotation of propeller and displacement of vane (Gill)
b Height 30 in, vane length 33 in, propeller 9 in diam.
c 6 lb
d Analog voltage proportional to vane angle and proportional to wind speed
e 0–70 miles/hour; 0°–360°azimuth; 50°–0°–50°elevation
f 3.1 ft distance constant and delay distance (approx 2 cps)
g ± 1% infinite resolution (azimuth potentiometer has 5% deadband)
h −30° C to +70° C
i 1000 ft (9 conductors)
j $685

4 Power supply (regulated) required for potentiometer excitation voltage

5
a 0–1 mA (100 ohm)
b Millivolt or galvanometer type (0–100m
or (0–1 mA & 0–5 mA galvo)

6 Balanced vane assembly with propeller o
front for wind speed measurement. Sem
portable – requires regulated DC power supply or regulated battery supply for vane potentiometers. Wind speed is self-generating with miniature tachometer generator. Used for short term three-dimensional wind measurements such as turbulence studies but subject to vane unbalance during periods of dew formati
or precipitation.

7
a National Bureau of Standards,
Radiophysics Laboratory,
Boulder, Colorado, U.S.A.
b Meteorological Branch, Department of Transport,
Toronto, Ontario, Canada.
c Los Alamos Scientific Laboratory,
Los Alamos, New Mexico, U.S.A.

See page xii for Key

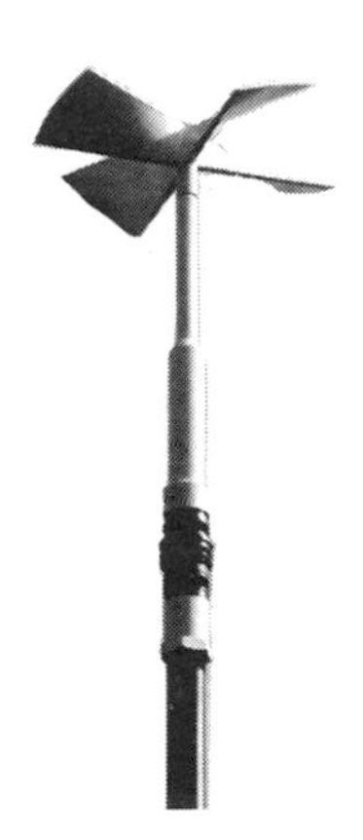

1 Propeller anemometer (27100)

2 R.M. Young Company,*
42 Enterprise Drive,
Ann Arbor, Michigan 48103,
U.S.A.

3
a Rotation of lightweight propeller (Gill)
b Propeller 9 in diam.; sensor 18 in long x 1 in diam.
c 15 ounces
d DC analog voltage – 500 mV at 21.3 mph
e 0–70 mph head into wind; 0–50 mph all angles
f 3.1 ft distance constant (approx. 2 cps)
g ± 1% cosine conformity 10% infinite resolution
h –30° C to +70° C; 0 – 100% rh
i 1000 feet (2 conductors)
j $148

4 Meter, recorder, or data logger

5
a 0–1 mA or 500–0–500μA
b Millivolt or 0–5 mA galvanometer type

6 Measures forward and reverse airflow and can therefore be used for vertical velocities, flux estimates, etc. Propeller rotates 1 revolution per foot of air driving miniature tachometer generator; polarity of signal reverses for charge of direction. Propeller follows cosine law within 10%

7
a C.S.I.R.O. Meteorological Physics Division, Melbourne, Australia.
b N.A.S.A., G.C. Marshall Space Flight Center, Huntsville, Alabama, U.S.A.
c Kennedy Space Flight Center, Florida, U.S.A.
d Air Force – Cambridge Research, Laboratories, Cambridge, Massachusetts, U.S.A.

See page xii for Key

1 Propeller vane (35001)

2 R.M. Young Company,*
42 Enterprise Drive,
Ann Arbor, Michigan 48103, U.S.A.

3
a Rotation of propeller and displacement of vane (Gill)
b Height 21 in, vane length 34 in
c 6 lb
d Analog voltages (2 channels)
e 0–90 mph/0°–360° (5% deadband in potentiometer)
f 3.1 feet distance constant; 2.9 feet delay distance
g ±1%, infinite resolution
h −30° C to +70° C; 0–100% rh
i 1000 feet (9 conductor)
j $354

4 Regulated power supply required for potentiometer excitation voltage

5
a 0–1 mA (100 ohm)
b Millivolt or galvanometer type (0–100 mV or 0–5 mA & 0–1 mA galvos)

6 Semi-portable – requires regulated DC power supply or regulated battery suppl for vane potentiometer. Wind speed sign is self generating using miniature DC tachometer generator. Threshold below 0.5 mph. Follows rapid fluctuations of wind speed.

7
a White Sands Missile Range,
Atmospheric Science Group,
White Sands, New Mexico, U.S.A.
b Florida State University,
Department of Meteorology,
Tallahassee, Florida, U.S.A.

See page xii for Key

1 Bivane (Bi-directional wind vane) (17002)

2 R.M. Young Company,*
42 Enterprise Drive,
Ann Arbor, Michigan 48103, U.S.A.

3
a Displacement of lightweight vane (Gill)
b Height 30 in, length of vane 28 in
c 6lb
d Analog voltage from linear potentiometers
e 0°–360° azimuth (5% deadband) & 50°–0°–50° elevation
f Delay distance 3.2 feet (approx. 2 cps)
g ±1%, infinite resolution
h −30° C to +70° C; 0 to 100% rh
i 1000 feet (9 conductor)
j $476 (U.S.)

4 Regulated power supply or battery for potentiometers

5
a 0–1 ma (100 ohms)
b Millivolt or Galvo (0–100 mV or 0–1 mA)

6 Measures azimuth angle and elevation angle. Wind direction only. Subject to imbalance during dew and precipitation. Fast response, low inertia, minimum overshoot.

7
a Department of Transport,
Meteorological Branch,
Toronto, Ontario, Canada.
b Department of Meteorology and Oceanography, University of Michigan,
Ann Arbor, Michigan, U.S.A.

See page xii for Key

1 Anemometer and direction vane (12301)

2 R.M. Young Company,*
42 Enterprise Drive,
Ann Arbor, Michigan 48103, U.S.A.

3
- a Rotation of cups and displacement of vane (Gill)
- b Vane 16 in high x 29 in long cup wheel 6 in diam., cups 2.4 in hemispheres (alum.)
- c Vane 2½lbs; anemometer ½lb
- d Analog voltages from generator & potentiometer
- e 0–100 mph & 0°–360° (5% deadband)
- f Delay distance 3.1 ft; distance constant 8.9 ft
- g ±1%. infinite resolution
- h −30° C to +70° C, 0–100% rh
- i 1000ft (7 conductor)
- j $329

4 Regulated power supply required for potentiometer

5
- a 0–1 mA (100 ohms)
- b Millivolt or galvo type (0–100 mV or 0–5 mA & 0–1 mA)

6 Rugged but sensitive matched vane and cup anemometer for horizontal wind measurements. Cup anemometer is useful for remote applications when line power is not available.

7
- a Department of Transport, Meteorological Branch, Toronto, Ontario, Canada.
- b Atmospheric Science Group, University of Texas, Austin, Texas, U.S.A.
- c Sea-Air Interaction Laboratory, E.S.S.A., Miami, Florida, U.S.A.

See page xii for Key

8
Water Content Meters

Many types of instrument have been designed for measuring the water content of porous materials such as soil. **Neutron scattering** is the most reproducible, accurate and precise method now available. It depends on an almost linear relationship between volumetric water content and the flux of slow neutrons produced when fast neutrons from a radioactive source such as amerycium-beryllium collide with hydrogen molecules. The density of soil can be measured **in situ** by the attenuation of gamma radiation. Commercial instruments are available for measuring water content alone with a neutron source and detector or water content and density together with **neutron** and **gamma-ray detectors**. Readings with these instruments give the average water content within the range of the scattered neutrons, usually about 10 cm in wet and 30 cm in dry soil. Different types of source-detector geometry are therefore needed for measurements at the soil surface and at depth. Changes in the distribution of soil water can also be estimated with tensiometers or with resistance blocks which are much cheaper and simpler to use than neutron meters but are inherently less stable, less accurate and less versatile. They are included in the references but not in the list of commercial equipment.

A number of instruments have been marketed for measuring the deposition of **dew** or the **wetness** of surfaces exposed to dew and rain. They provide a useful guide to the period for which natural surfaces remain wet. The amount of condensation depends on the heat balance of the surface and will rarely be equal on vegetation and on the artificial surface of a dew gauge.

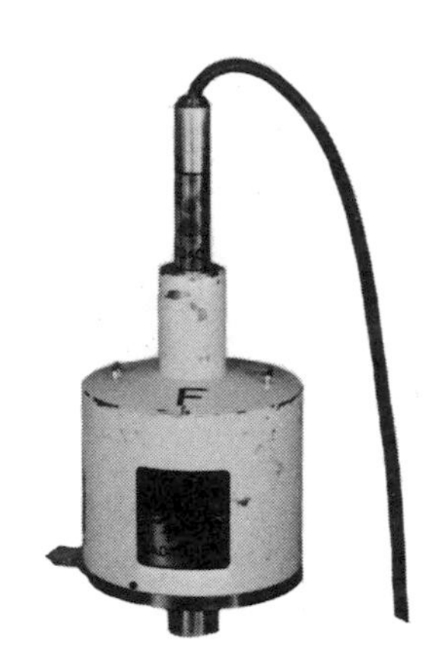

1 Neutron moisture and gamma density meter (type LB 362)

2 Laboratorium Professor Dr. Berthold,* 7547 Wildbad Calmbacher Strasse 22, Germany (BRD).

3
a Neutron and gamma back-scattering
b Surface source 14 x 35 x 34 cm tall depth probe 36 mm diam., 66 cm long; shield 23 cm diam.
c Surface sources 15 kg; depth probe 4 kg; shield 28 kg; ratemeter 4 kg
d Ratemeter
e 0 to 40% moisture by volume, 1.2 to 2.5 g cm^{-3} density
f 5, 10, 20 sec
g $\pm$1% by weight
h -20° to 50°C
i 3 m standard

4 Battery charger

5
a Meter displays density and moisture
b Reorder output available on request

6 Surface sources (left hand photograph) can be used for average moisture content or density in range 0 to 30 cm depth; pro (right hand photograph) is used for 30 cm to 40 cm depth. 12 hours operation from rechargeable batteries.

See page xii for Key

1 Neutron moisture meter

2 A/S Danbridge,
47 Brigadavej,
Copenhagen 5, Denmark.

3
a Neutron scattering
b Probe 38 mm diam.
c Probe 2kg, shield 11 kg, scaler 14kg
d Scaler or ratemeter
e £1400; ratemeter £110 extra

7 See Bell and McCulloch (1969), p. 252

See page xii for Key

1 Dew recorder Kessler type (64a)

2 R. Fuess,*
D–1 Berlin 41,
Dunther Str. 8,
Postfach 350, Germany (BRD).

3
a Measuring of dew on artificial surface
b 285 mm length x 130 depth x 190 mm height
c 3.4 kg
d Chart
e 0 to 6 g quantity of dew, one chart division is 0.1 g
j £95

5
b Pen records on chart drum which rotates at 11 mm/hr or 40 mm/day, clockwork drive

6 Oil damped balance system

7 Meteorological stations all over the world

See page xii for Key

Surface wetness recorder Woelfle-Type
(64b)

R. Fuess,*
D–1 Berlin 41,
Dunther Str. 8,
Postfach 350, Germany (BRD).

Contraction of hemp threads
50 x 14 x 17 cm
2.8 kg
Chart
0 to 50 arbitrary units
£70

Pen records on chart drum rotating in
1 day or 1 week; clockwork drive

Meteorological stations all over the world

See page xii for Key

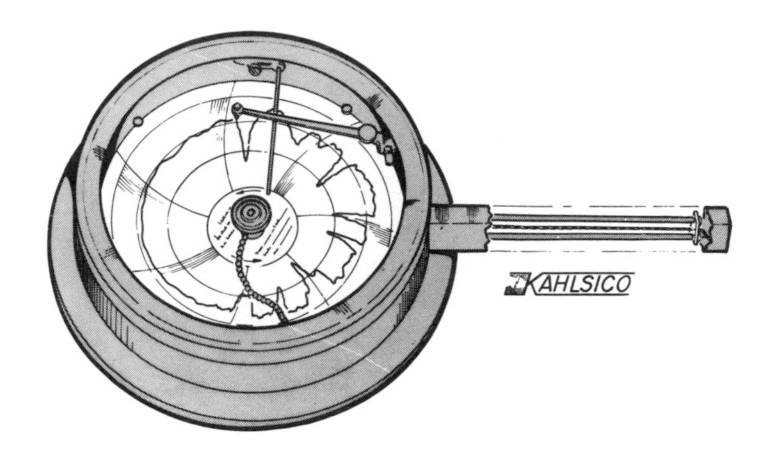

1 Wetness recorder

2 Kahl Scientific Instrument Corporation,
P.O. Box 1166,
El Cajon,
California 92022, U.S.A.

3
a Contraction of fibres
b Sensor 2 mm diam. x 15 cm long; housing 26 cm x 8 cm
c 3.7 kg
d Chart

5
a Pen records on 22cm diam. chart rotating in 1 week; clockwork drive.

6 A more elaborate version of this instrument has a bimetallic element for recording temperature on the same drum; range –20° to +40°C ± 1° C.

See page xii for Key

1 Dew recorder (M 35)

2 Mashpriborintorg,
Smolenskaya-Sennaya Square, 32/34,
Moscow G-200,
U.S.S.R.

3
a Change in weight of plate
b 46 x 27 x 22 cm
c 6kg
d Chart record
e 0.3 to 4.8 g equivalent to 0.02 to 0.3 mm dew
g Response threshold < 0.2 g
h 0°to 40° C

5
a Clockwork driven chart, 26 hours per revolution

See page xii for Key

1 Neutron moisture and gamma density meter

2 N.E.A.,*
39-43 Industriparken,
DK 2750 – Ballerup,
Denmark.

3
a Neutron and gamma back-scattering
b Neutron and gamma probe 38 mm diam. 58 cm long; shield 17 cm diam. 60 cm long
c Probe and shield 10 kg
scaler/ratemeter 11 kg
d ratemeter and scaler
e 0 – 100% water content
f 30 sec
g ± 1% water content
h −10°to +50° C
i 100 m
j £1800

4 None

6 Fully portable with built-in accumulator

7
a Users in 16 countries, e.g. I.A.E.A., Vien
Austria; F.A.O. Rome, Italy; Daniel Ator
Energy Commission.

(see Bell and McCulloch, 1969, p. 252)

See page xii for Key

1 Wallingford soil moisture meter

2 D.A. Pitman Ltd.,*
Mill Works, Jessamy Road,
Weybridge, Surrey, U.K.

3
a Neutron scattering
b Probe 3.8 cm diam. x 74 cm; transport shield & meter 15 cm diam. x 104 cm
c 8 kg (probe alone 1.7 kg)
d Probe produces 2Vm 2 μ sec negative pulses
e Full range of soil moisture
f Depends on precision required
g 1% moisture volume fraction
h −10°C to +40°C
i No reasonable limit
j Probe £850 (ratemeter £150) (ratescaler £300)

4 Power supply integral with ratemeter or ratescaler

5
a Ratescaler – reads count rate direct over 16 sec or 64 sec period, has binary coded digital output option; ratemeter has analogue output voltage facility
b 0 – 5V to monitor ratemeter output

6 Probe can be supplied alone or with either integral or separate ratemeter or ratescaler. Complete assembly is easily portable, rugged and very stable with temperature and time.

7
a Institute of Hydrology,
Howbery Park,
Wallingford, Berkshire, U.K.
b National Vegetable Research Station
Wellesbourne,
Warwickshire, U.K.
c Dept. of Physiology and Environmental Studies,
University of Nottingham,
Sutton Bonington,
Loughborough, Leicestershire, U.K.
(see Bell and McCulloch, 1969, p. 252).

See page xii for Key

1 Neutron moisture meter (1257)

2 Troxler Laboratories Inc.,
P.O. Box 5536,
Raleigh, N.C. 27607, U.S.A.

3
a Neutron scattering
b Probe 38 or 47.4 mm diam.
c Probe 1.4 kg, shield 5 kg, scaler 10.4 kg
d Ratemeter or scaler
e 0 to 35% vol water
i 300 ft
j Probe and scaler $4400
probe and ratemeter $3000

6 Surface adaptor available.
A wide range of moisture content and density probes is manufactured by this firm (see Bell and McCulloch, 1969, p. 252)

See page xii for Key

1 Surface moisture and density meter (HDM 4)

2 Viatec Pty, Ltd.,*
P.O. Box 23,
Plumstead,
Cape, S. Africa.

3
a Neutron scattering
b Scaler 9½ in long by 8 in wide by 11¼ in high
c 75 lbs in transit case
d Analogue
e Moisture 0 to 50 lb/ft^3
Density 80 to 180 lb/ft^3 (approx)
f 100 sec
g Crystal controlled timing with 0.01% accuracy
h From –20° C to +65° C
i Up to 1000 ft
j $5,500

4 Either AC 115/220/250 volts drawing **10** milliamps at 50 – 60 Hz, or DC 6 to 12 volts drawing 180 milliamps

5 Meter

6 The HDM 4 is designed to measure, rapidly and non-destructively the in-place density and moisture of soils in road building materials. This system is completely portable and rugged. The transistorized unit has a very low power consumption

7
a Broken Hill Proprietary Co. Ltd.

b Australian Iron and Steel

See page xii for Key

9

Water Vapour Flux Meters

The flux of water vapour in the atmosphere is usually derived by combining measurements of net radiation, with temperature, humidity and wind gradients using equipment reviewed in previous sections. The flux can also be determined by correlating fluctuations of humidity and vertical velocity and systems of this type will be found among the references. A few commercial systems are available but have not been included because they are tools for investigating atmospheric turbulence rather than evaporation. The evaporation from vegetation can be measured directly with a lysimeter (which measures 'looseness' i.e. drainage) or weighing machine but these instruments are usually custom-built and are not offered by manufacturers. Evaporimeters in the form of pans, wet filter paper or wet ceramic plates are available commercially but have not been included in this survey. They are climatological instruments and their readings have no place in micrometeorological analysis.

Instruments are available for measuring the rate of evaporation from leaves or from skin. A **diffusion porometer** is an instrument for measuring the resistance of stomatal pores on a leaf from the rate of transpiration into a small cup containing a humidity sensor. A **sweat meter** (or sudorimeter) measures the increase in water vapour content of air passing through a cup attached to the skin surface.

1 Sweat meter (Custance Sudorimeter)

2 Canadian Research Institute,
85 Curlew Drive,
Don Mills,
Ontario, Canada.

3
a Measurement of evaporation rate
b Sensor 4 x 3 x 2 cm
c Sensor 20 g
d Digital signal and analogue voltage
f 0.2 sec
g Resolution 0.1% rh
h −40°to +40° C
i 20 m
j $1850

4 Mains 115V or 220-240 V,
50 Hz. Dry air at 2 to 25 lb/in^2

5
a Digital readout incorporated
b Milliammeter

6 A sensor system maintains a constant rh c
about 45% in a small cup attached to the skin. Mainly used for human studies but 'the ingenuity of the researcher is the onl limiting factor'. A completely portable system is being designed.
(see Custance and Laflamme, 1961, p. 25

See page xii for Key

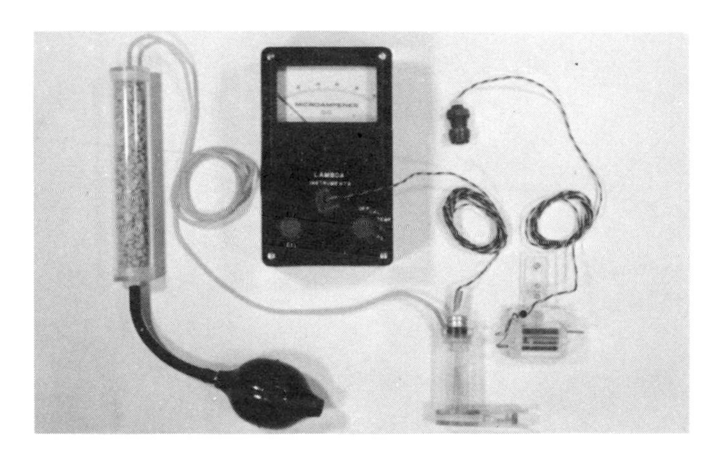

1 Stomatal diffusion porometer

2 Instrumentation Systems Center,*
University of Wisconsin,
1500 Johnson Drive,
Madison, W1 53706, U.S.A.

3
a Increase of humidity in porometer cup
b Sensor 1 in diam. x 3 in long; control unit 7 x 5 x 3 in
c control unit 1 lb
d Current
e Appropriate for stomatal resistance
g Repeatability about 0.3 sec/cm
h 40°to 90° F
Control unit $190
probe $90

4 Internal batteries 13.5 V

5
a Meter incorporated

6 (see Kanemasu, Thurtell and Tanner, 1969, p. 254)

See page xii for Key

10

General Purpose Recorders

o eliminate a very large number of mains or line-operated recorders, the survey 'as confined to battery- or clockwork- driven instruments. These include **otentiometric recorders** (measuring voltage) in which a DC input signal is nplified before being fed to a servo-operated null-balance system; galvanoeters (measuring current) with or without an amplifier; and Wheatstone ridges (measuring resistance) in which the output signal can be the voltage cross an unbalanced bridge or the displacement of a slide wire to achieve alance. Single-channel potentiometers and some galvanometers use pen, ink nd conventional paper or a stylus and pressure sensitive paper. Other galvanoeters record intermittently by using a **chopper bar.** One type of **event recorder** a simplified form of magnetic tape recorder in which the time of a series of vents is stored on tape. Magnetic tape recorders have the advantage of portaility and battery operation but some commercial models have proved unreable in field conditions. Larger and more reliable data loggers incorporate igital voltmeters but need a stable mains power supply. Output can be recorded n a strip printer, typewriter, paper tape or magnetic tape. Models and ariations were too numerous to include in the survey.

1 Magnetic tape recorder

2 d-Mac Ltd.,*
Queen Elizabeth Avenue,
Hillington,
Glasgow, S.W. 2, U.K.

3
a Analogue to frequency conversion
b 39 x 35 x 15 cm
c 11.5 kg
d Direct 0 to –5V DC input impedance 130 kohms; with x 500 amplifier, input impedance 10 kohms;
e ± 1% from -10 to 36°C
g ¼in tape, 2 in/sec, 500 B.P.I., recording time 0.3 sec
h maximum 10, 1 to 60 minutes
i 13.5 + 1.5V, 350mA

4 The basic unit can be supplied with a range of plug-in units converting signals from different types of transducer to 0 to –5 Volts. This corresponds to a pulse count from 00 to 99. Translation unit available.

See page xii for Key

Battery operated recorder

EKO Instruments Trading Co. Ltd.,*
2-1 Ohtemachi 2-chome,
Chiyoda-ku, Tokyo 100, Japan.

Galvanometer with chopper bar
14.5 x 15.0 x 24.0 cm
5.5 kg
0 – 25mV
± 0.5%
70 mm, 20 mm/hr
1 channel, 1 minute
1.5V DC
–10°to 40° C
$280

The recorder can be used in combination with a thermopile-type solarimeter. A dry battery lasts 45 days and the chart 27 days. No ink is required. The standard model of this recorder is designed for indoor use only.

See page xii for Key

1 Incremental magnetic tape recorder

2 Epsylon Industries Ltd.,
Faggs Road, Feltham,
Middlesex, U.K.

3
a Analogue to digital conversion
c 5.5 kg
d 100 mV fs at 100 k ohms
e Each scan includes logging of two standard voltages
g ½ in tape
h 12, 5 to 60 minutes
i 12V DC
j −30° to 40° C

4 Recording as 8-bit binary parallel word at 100 words per in. on 600 ft single-play tapes. Clock track and block markers provided. Playback units available.

See page xii for Key

Battery operated recorder (A 601 C)

Esterline Angus, Division of Esterline Corp.,*
P.O. Box 24000,
Indianapolis, Indiana 46224, U.S.A.

Permanent-magnet moving-coil ammeter with solid state operational amplifier.
14 x 8 x 9 in
34 lb
0-100 mV and 0-1 mA DC
1% of full scale
6 in wide ¾ in/min to 12 in/hr
1 or 2 channels, 1/2 second response
9V DC
+20°to +120° F
$625

Continuous operation for 8 days

See page xii for Key

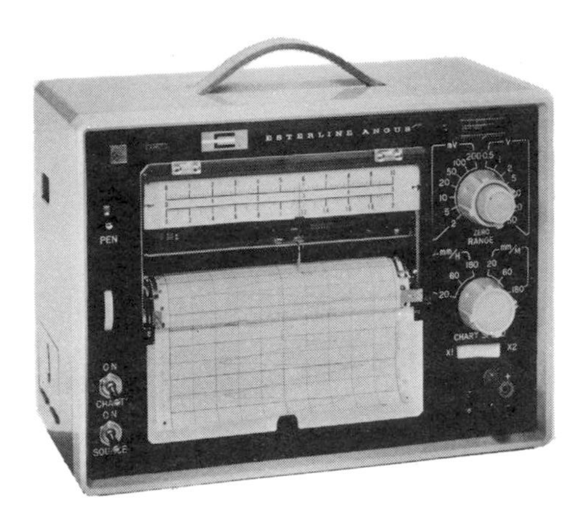

1 Battery operated recorder (T 171 B)

2 Esterline Angus, Division of Esterline Corp.,*
P.O. Box 24000,
Indianapolis, Indiana 46224, U.S.A.

3

a Millivolt potentiometer with servosystem
b 13 x 10 x 6 in
c 15 lb
d 2 mV to 50V full scale, 10 k ohms source impedance
e $\pm$ 0.5% of span
g 150 mm chart; speeds of 20, 60 & 180 mm/min & hr with 2X multiplier
h 1 channel
i 120 V AC 50/60 Hz
7.5–9V DC internal, or 12V DC external
j +50°to +110° F
k $800

4 Can use rechargeable nickel-cadmium batteries. Battery charger built in 12 hours on rechargeable batteries. 8 hours on 1½V 'D' cell. 24 hours on alkaline 'D' cell. Up to one week on external 74 amp. hr 12V auto battery. Model T. 171 B. Portable case.

See page xii for Ke

1 Battery operated recorder (A 601 C)

2 Esterline Angus, Division of Esterline Corp.,*
P.O. Box 24000,
Indianapolis, Indiana 46224, U.S.A.

3
a Permanent-magnet moving-coil measuring element with solid state operational amplifier
b 14 x 8 x 9 in
c 34 lb
d 0 to 100 mV and 0 to 1 mA DC
e ± 1% fs
g 6 in wide; ¾ in per min to 12 in per hour
h 1 or 2 channels, 1/2 second response
i 9 V DC (internal)
j +20°to +120° F
k $625

4 Continuous operation for 8 days

See page xii for Key

1 Recording galvanometer

2 Joens, *
Dusselforf, Germany (BRD).

3

a Chopper bar galvanometer
b 300 x 200 x 300 mm
c 12 kg
d 100mA full scale deflection
e 1%
f Good stability when horizontal
g Chart width 120 mm, speed a few cm/hour
h 6 channels cycle time, 6 x 60 sec
i Chart and sampler are clock driven. The spring can be wound up by hand. Electronic circuits for several sensors are powered from Deac cells.
j 0°– 50° C, 20 – 100% rh
k £290 without electronic circuits and sensors

5 For field use electronic circuits are built in the recorder by T.F.D.L., Wageningen, Netherlands to measure temperature rh, conductivity and pH. Light sensors and, for wind velocity cup anemometers, can be recorded without electronic circuits.

See page xii for Key

1 Magnetic tape event recorder

2 Lintronic Ltd.,
54 Bartholemew Close,
London, E.C.1., U.K.

3
a Analogue to digital conversion
b Watertight case 10 x 8 x 8 in
c 9 lb
d max. input rate, 5 events per sec
e clock accuracy ± 2 min per week
g ¼ in magnetic tape, 0.01 in per step
h 3 event channels, 1 time marker
i 9 to 15V DC, 10mA at 2 events/sec

4 Designed as tipping bucket rainfall recorder but can be used with any contact closure device provided 3d is not exceeded.

See page xii for Key

1 Magnetic tape recorder

2 Microdata Ltd.,
Allweather House,
High Street, Edgware,
Middlesex, U.K.

3

a Analogue to digital conversion
b 4 x 6 x 11 in
c 6 lb
d 0 to 4 volts
e 0.5% fs
h 12; sample time variable
from 5 min to 1hr
i 5 – 8 V DC
j −10°to +40° C
k £400

4 Uses standard 0.15 in x 600 ft
cassette, taking 114,000 measurements

See page xii for Key

1 Magnetic tape recorder

2 Normalair – Garrett.

3

a Analogue to digital conversion
b 8 x 16 x 9 in
c 24 lb + battery
d 0 to 100 mV
e 0.1% resolution
h 8
i 6 V DC
k £1500

4 Uses ¼ in x 600 ft cassette taking 4000 scans

See page xii for Key

1 Magnetic tape recorder

2 Plessey Company Ltd.,*
Environmental Sensors Division,
Vicarage Lane, Ilford, U.K.

3
a Analogue to digital conversion
b 38 x 18 x 17 cm
c 4.5 kg
d 0 to 5V or 1 to 20 kohms
e ± 0.2%
g ¼ in magnetic tape; 8 sec per channel
1, 2 or 4 scans per hr
h 8; 70 sec approx.
i Battery operated
±12V DC, 190 mA
j Tropicalised: –10° to +60° C
Standard: –10° to +40° C other ranges
to order
k £630 (logger only)
£1685 (complete with sensors)

4 Equipment ranges from basic logger to complete systems including meteorological and/or water quality sensors. Period of unattended operation in field normally up to 2 months. Translator unit available.

See page xii for Key

1 Incremental magnetic tape recorder

2 Rapco Electronics,*
10 Joule Road,
Basingstoke,
Hampshire, U.K.

3
a Analogue to digital conversion
b 17 x 14 x 10 in
c 40 lb
d ± 100 mV fs
d Accuracy ± 1% of fs; resolution ± 0.1% of fs
f Accuracy specification includes temperature drift
g ¼ in magnetic tape, incremental drive
h 10 analogue channels, 1 digital channel
i Self-contained rechargeable batteries
j 0°C to 50°C
k £1,200

4
a Operating time (unattended) in field typically 1 month, can extend up to three months in same mode of operation
b Unit is contained in weatherproof, sealed light-alloy case
c Scanning intervals (which can range from 1 per sec to 1 per hr) are generated by an internal crystal clock having an error of only ± 5 min/month

See page xii for Key

1 Digital event recorder

2 Rauchfuss Instruments and Staff Pty. Ltd.,
11 Florence St.,
Burwood,
Victoria, Australia 3125.

3

b 17 in diam. x 18 in high
c 27 kg
g 5 hole paper tape
h 3 or 6
i Battery
7.5V DC triple-charged by silicon solar cell

4 A wide range of sensors and transducers is available to measure wind speed, humidity, temperature, radiation, sunshine duration, rainfall, water level, etc.

See page xii for Key

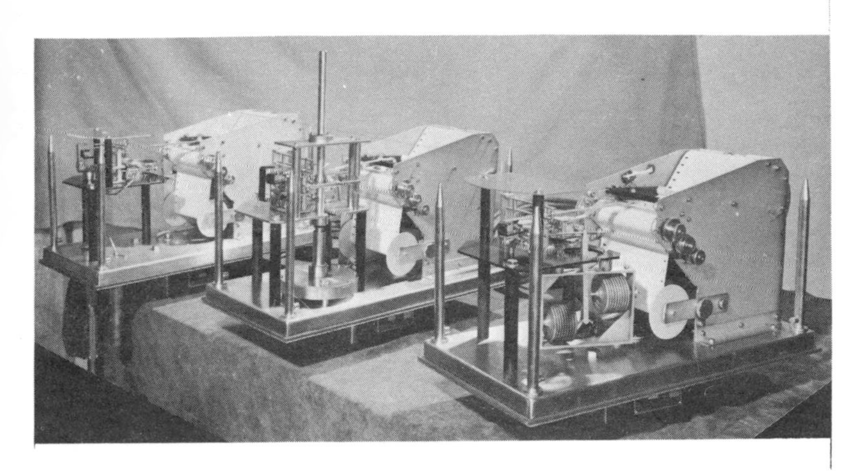

1 Battery operated strip chart recorder
(Sumner Mark II)

2 Rauchfuss Instruments and Staff Pty. Ltd.,
11 Florence St.,
Burwood,
Victoria, Australia 3125.

3
a Galvanometer and chopper bar
b 20 x 9 x 13 in
c 32 lb
e ± 1% fs
g 4 in; 4 in in 24 hr
h 1 or 2
i 135V DC

4 A wide range of sensors is available
for measurements of water level,
water pressure, temperature,
humidity, rainfall, wind
direction and speed.

See page xii for Key

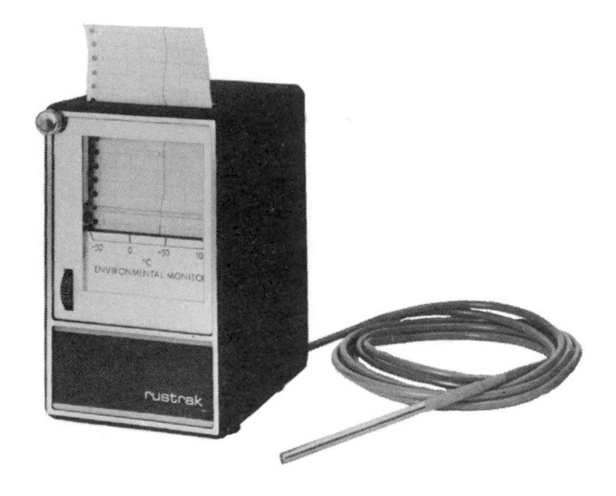

1 Battery operated recorder (Model 2194)

2 Rustrak Instrument Division,*
Municipal Airport,
Manchester, N.H. 03103, U.S.A.

3

a Galvanometer with chopper bar
b 6 x 4 x 5 in
c 3 lb
d 1μA or 0 – 10 mV with max. source impedance 100 ohm
e ± 2%
f ± 2%
g 2 5/16 in; 1/8 in to 50 ft per min
h 1 channel
i 12V DC, 30mA
j 0° to 50° C
k $293

4 Can operate unattended from 2½ years to 10 days depending on speed and size of batteries used. Similar recorders (as illustrated) can be supplied with a wide range of thermocouples, thermistors or nickel wire sensors.

See page xii for Key

1 Battery operated chart recorder

2 Rustrak Instrument Division,
Municipal Airport,
Manchester, N.H. 03103, U.S.A.

3
a Null-balance potentiometer
b 4 x 6 x 6 in
c 4 lb
d 0 to 100mV; 0 to 1mV with plug-in amplifier
e ± 0.5% fs
f ± 0.025μV/°C (0° to 50°C)
g 2 5/16 in; ½ in/hr to 4 in/min
h 1 channel
i 12V DC; 50mA
j 0° to 50° C
k Basic unit \$299; amplifiers \$75 to \$170
DC models add \$60 to basic cost.

4 Can operate up to 1 month unattended. For use with I/C, C/A, C/C or platinum thermocouples with P 4003 amplifier

See page xii for Key

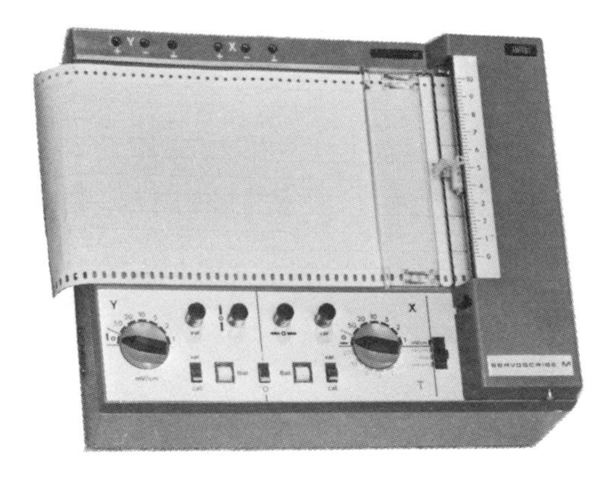

1 Battery operated xy or yt recorder (Servoscribe M)

2 Smith Industries Ltd.,
Kelvin House,
Wembley Park,
Middlesex, U.K.

3

a Null balance potentiometer
b 22 x 26 x 11 cm
c 2.8 kg
d 1, 2, 5, 10, 20 or 50 mV/cm
e ± 1% fs, ± 0.5% dead band
f < 0.2%/10°C from –10°to +50°C
g y axis 10 cm, x axis 15 cm; 1 to 20 min/cm
h 2
i 12V DC

4 Accepts sources with resistance between 10 and 250 kohms

See page xii for Key

Digital recorder

T.F.D.L., Dr. S. L. Mansholtlaan 12,
Wageningen, Netherlands.*

Digital-clock controlled scanner, digital voltmeter and punch tape output.
2 boxes each 750 x 520 x 430 mm
50 kg
three digits
0.5% or 2 units of last digit
Good stability
Punch tape, 10 characters/sec
scanning speed 1 or 2 channels/sec
1-10 sets of 10 channels each; every 1, 5, 10, 15, 30, 60 min
Battery powered, 24 V DC 50 Ah
0°– 40° C; 20 – 100% rh
£4,600 without sensors

Many electronic sensors can be employed e.g. for temperature, light, rh, wind velocity, conductivity, pH, etc.
The recorder with 20 sensors and cycle time every hour will operate 1 month unattended in the field.

See page xii for Key

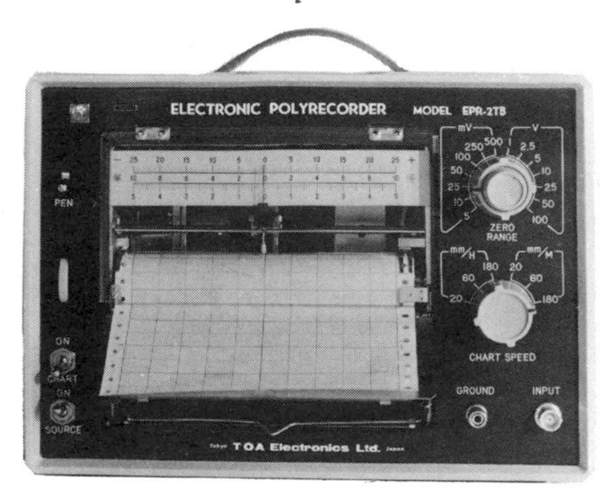

1 Battery operated recorder (EPR-2TB)

2 Toa Electronics Ltd., 235 Suwa-Cho, Shinjuku-Ku, Tokyo, Japan.

3

a Null balance potentiometer
b 32 x 23 x 16 cm
c 6 kg
d 14 switched ranges, fs from ± 5 mV to ± 100 V
e 0.5% fs
f Drift less than 50 μV/hr
g 15 cm; 20, 60, 180 mm per min and per hr
i 9V DC
k £280

4 Zero adjustable over full scale on all ranges

See page xli for Key

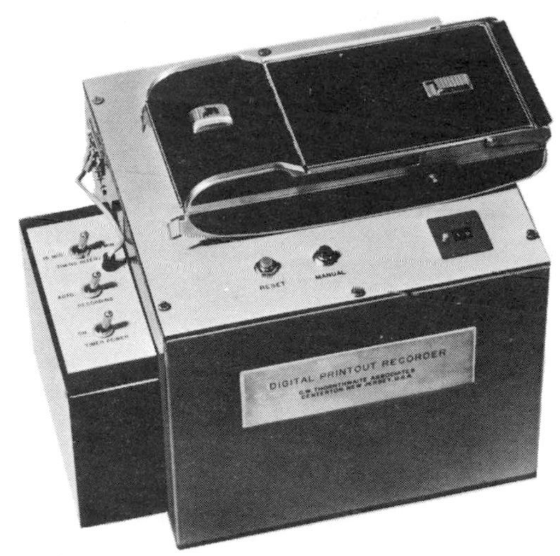

Battery operated counter recorder (Model 70a)

C.W. Thornthwaite Associates,*
Route 1, Centerton,
Elmer, New Jersey 08318, U.S.A.

Photography of electromagnetic counters
8 x 10 x 10 in plus a camera protrusion of 3 in
17 lb including camera and precision timer
0 to 99999 counts per recording period
± 1 count per recording period
stable in operation through count rates of
0 to 10 counts/sec
the registers are reproduced on a 3 x 4 in
print
9 channels with switch selection of 15 min
or 60 min recording intervals
12 V DC, mean power consumption 60 mW
Recommended for use above 0°C
$1000

The instrument is designed for automatic recordings of one hour time-integrated averages of wind profiles (up to 9 levels). The recorder will operate unattended for 54 hours and then automatically shut itself off until reset by the operator.

See page xii for Key

11
Integrators

Several firms now offer equipment for integrating a voltage or current with respect to time. These **integrators** are based on stable DC amplifiers with solid state components and on the time-constant of a resistance-capacity circuit. Most integrators accept signals of millivolt level, e.g. from radiometers, and display their output on a counter. Provision is usually made for changing ranges and for backing-off part of the input voltage. This section describes general-purpose integrators which can be used with any sensor giving an appropriate output. Integrators designed for specific temperature and radiation sensors will be found under sections 1 and 3 respectively.

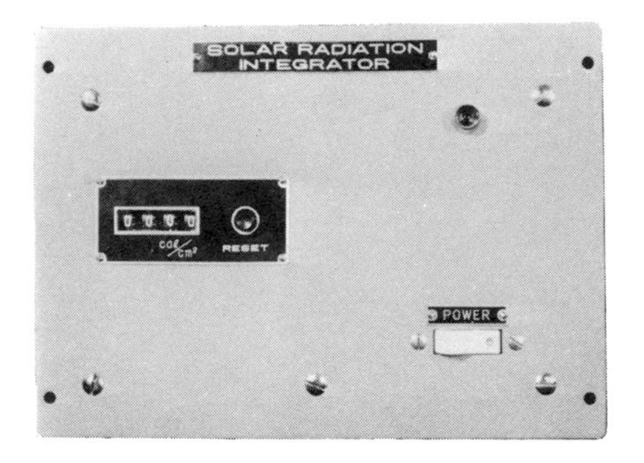

1 Integrator (SR1 520)

2 Iio Denki Ltd. *
Yoyogi 2-27-18, Shibuya-ku,
Tokyo, Japan.

3
a Based on potentiometric mV meter
b 15 x 21 x 30 cm
c 7 kg
e ± 1 cal/cm^2 when used for solar radiation
g Counting interval 1 min
i 100 V AC; 50 or 60 Hz
j −10° to 50° C; 0 to 80% rh
k $700 to 810

5 Noshi Denshi solarimeter or Gorzyinsky solarimeter are used as sensors for this integrator

See page xii for Key

1 Integrator

2 EKO Instruments Trading Co. Ltd. *
2-1 Ohtemachi 20-chome,
Chiyoda-ku, Tokyo 100, Japan.

3
b 36.5 x 26.5 x 24 cm
c 15 kg
d 0 to 2.0 cal cm^{-2} min^{-1}
e 1.5%, 0.5 mV
g 1 printing per hour
h 1 channel
i 100-220 V AC; 50 or 60 Hz
j −10° to 40°C
k $1,530

4 The standard type of this recorder for use in laboratory conditions

See page xii for Key

1 Millivolt integrator

2 Instrumentation Systems Center, *
University of Wisconsin,
1500 Johnson Drive,
Madison, W1. 53706, U.S.A.

3
b 8 x 6 x 8 in
c 2 lb
d ± 50 μV to 5 mV in 1 range
e 50 to 200 μV, ± 3%; 200 μV to 5 mV, ± 1%
f Minimum of 4 hr for full scale drift with input shorted
i 115 V 60 Hz or internal ±15 V DC, 40 mA
j 40° to 90° F
k $580

See page xii for Key

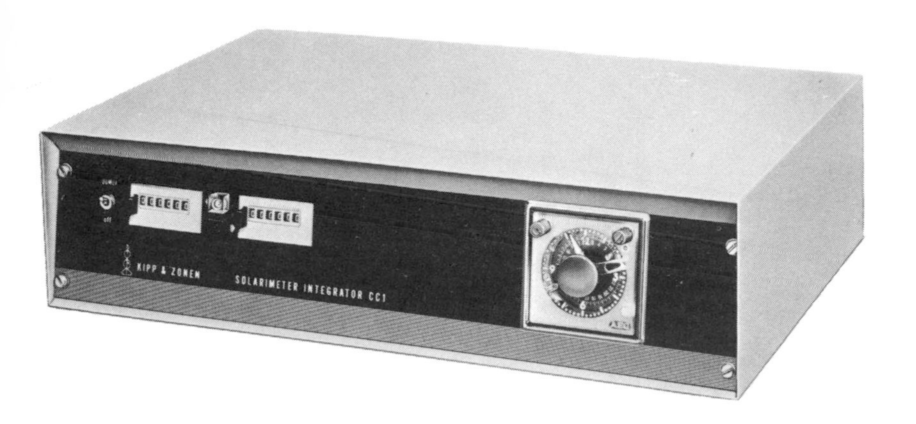

1 Integrator (CC1)

2 Kipp and Zonen,
P.O. Box 7,
Delft,
Netherlands.

3
a Integration of slide-wire voltage
b 44 x 29 x 12 cm
c 10 kg
e 0.3%
h 1
i 115 or 220 V 50/60 Hz
k £250

4 Incorporates two counters switched by time-clock for measuring solar radiation over 24 hours, etc. Intended for use with potentiometric recorder fitted with re-transmitting slide wire.

See page xii for Key

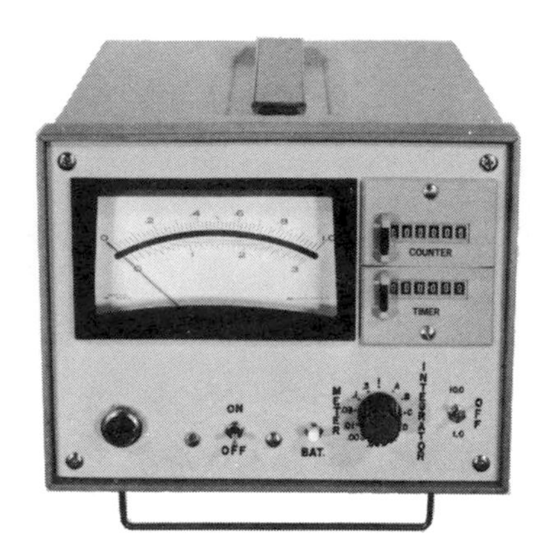

1 Electronic integrator

2 Lambda,
2933 North 36th,
Lincoln,
Nebraska 68504, U.S.A.

3
b $6\frac{1}{2}$ x 8 x 11 in
c 5 lb
d 10^{-3} to 1 μEinstein cm^{-2} sec^{-1} or voltage exceeding 1 mV
e 1% for integration between 1 and 10 counts per sec
i 115 V AC or battery
k $750

4 Primarily intended for use with photon meter (p. 136). Includes digital timer

See page xii for Key

1 Millivolt integrator

2 Lee-Dickens Ltd.,
69 Kettering Road,
Northampton,
Northamptonshire, U.K.

3

b 12 x 7 x 2.4 in
d 0 to 100 mv, 1 k ohm input
e ± 0.5% linearity better than 0.1%
f Better than 0.02% per °C
i 100/120 or 200/240 V AC
12 or 24 V DC (with inverter)

4 Zero suppression up to ± 25% fs, seven digit reset counter fitted as standard. Amplifier available for thermocouple outputs.

See page xii for Key

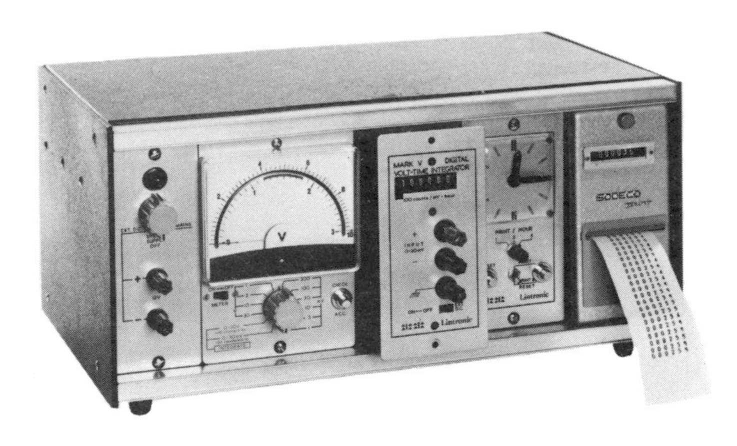

1 Millivolt integrator

2 Lintronic Ltd.,
54-58 Bartholomew Close,
London, E.C. 1., U.K.

3

a Differential feed-back chopper stabilized amplifier
b 12 x 15 x 6 cm
c 0.5 kg
d 0.1 mV to 50 mV; 5 μV to 5 mV
e 100 counts per mV hr; $\pm$ 0.5% absolute or $\pm$ 2 counts per hr
f Negligible long term drift of $\pm$10 μV @ 30° C
h One
i 110 or 230 V AC
15 V DC internal batteries
j – 10°C to 60°C
k £100

4 Input can be offset for bipolar signals.
Multi-channel versions available
with print-out counters – see photograph

See page xii for Key

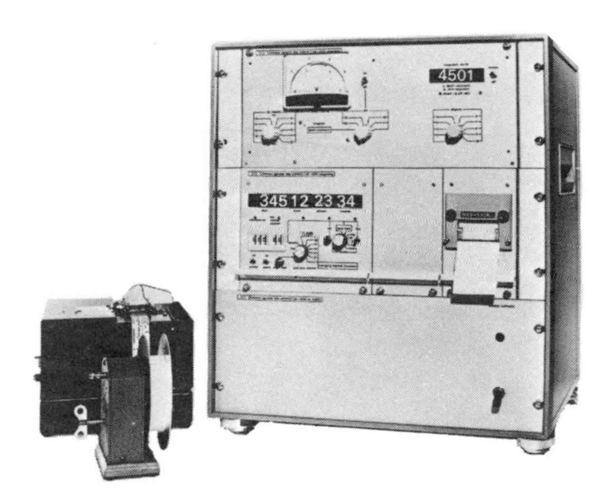

Multi-channel millivolt integrator

Lintronic Ltd.,*
54/58 Bartholomew Close,
London, E.C.1., U.K.

19 x 12 x 17 in
0 – 1, 10, 30 mV, input resistance 1 M ohm
± 0.5% reading
–0.03% per °C from –10°to +60° C
8 hole paper tape
10 channels standard; 5 to 30 min
print out
115 V 60 Hz, 230 V 50 Hz
28 VDC
£3,150

For voltage integration, one count can represent 12 or 36 mV sec. Input can be offset by ± 20 mV on each channel. Crystal controlled digital clock included. Input channel can be replaced by pulse count input. Magnetic tape recorder interface available.

See page xii for Key

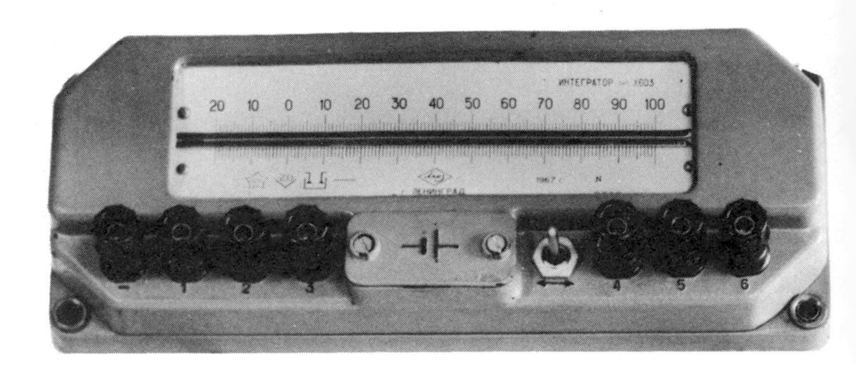

1 Electrolytic integrator

2 Mashpriborintorg,
Smolenskaya-Sennaya Square, 32/34,
Moscow G-200,
U.S.S.R.

3

a Coulometer measuring volume of hydrogen
b 200 x 80 x 54 mm
c 0.7 kg
d Various, e.g. 40 μAh fs
e ± 2.5% for 10°C
j +5° to +40°C; 0 to 80% rh

See page xii for Key

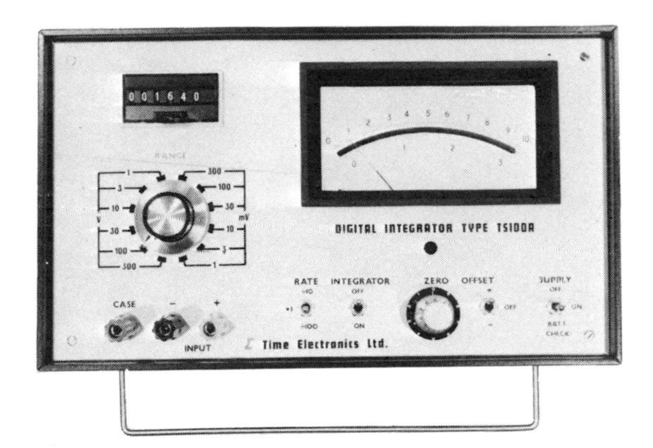

Millivolt integrator (TS 100)

Time Electronics Ltd.,
199a High Street,
Orpington,
Kent, U.K.

20 x 14 x 14 cm
3 kg
10 ranges, 1 mV to 30 V fs
± 1% absolute, threshold 0.1% fs
over-range 300% fs
< 0.5 μV/°C and 5 μV per day
One
£125

Can be offset to ± full scale on all ranges. Ramp output displayed on meter and integrated total on 6 digit counter. Two count rates available; the more precise is 33.3 μV sec per count. BCD output available in place of standard counter.
Also available in modular form or with strip printer, etc.

See page xii for Key

12

Telemetry Systems

ne of the first principles of micrometeorological measurement is that neither he instrument nor the observer should disturb the environment which is being neasured. Sensors are therefore kept as small as possible and are designed to ive an electrical output for transmission to a remote reading or recording evice, a process known as **telemetry**. Most of the instruments described in this andbook are intended for connection to a recorder by a cable and maximum able lengths are quoted under 3i. In animal and human studies, however, ables impose a serious restriction to movement and radio links are sometimes sed to transmit signals from sensors to recorders. This process is known as **adio-telemetry**. The versatility of radio-telemetry has been greatly increased y the use of miniature components and integrated circuits which allow transnitters to be attached even to small animals without interfering with normal ehaviour.

A number of firms now manufacture telemetry equipment using frequencynodulated V.H.F. signals. The suitability of a particular system for microneteorological studies depends on factors such as the size, portability and ffective range of the transmitter and the life of the power pack. The potential ser of radiotelemetric equipment is therefore recommended to investigate the he wide range of systems now available commercially and to seek expert advice n selection. As it proved impossible to provide a brief and adequate description of radio telemetry equipment within the format of this handbook, a list of nanufacturers is included instead.

Manufacturers of Radiotelemetric Equipment

Beckman Instruments Inc.,
Spinco Division, 1117 California Avenue, Palo Alto, California 94304, U.S.A.

Biotronics Incorporated,
838 Butte Street, Redding, California, U.S.A.

Cinortcele,
1 Lakeland Drive, Frimley, Camberley, Surrey, U.K.

Dynatel Data Systems Ltd.,
Spur Road, North Feltham Trading Estate, Feltham, Middlesex, U.K.

Elther N.V.,
66 Vaartweg, Hilversum, The Netherlands.

Ether Engineering Ltd.,
Park Avenue, Bushey, Hertfordshire, U.K.

Hayakawa Electric Co. Ltd.,
1-232 Nishitanabe, Abeno-Ku, Osaka, Japan.

Fritz Hellige & Co. GMBH,
78 Freiburg im Breisgau, Heinrich-von-Stephan-Strasse 4, Germany (BRD).

Messerschmidt-Bölkow-Blohm GMBH,
Ottobrunn bei München, München 8, Germany (BRD).

Mitsubishi Electric Corp.,
2-12 Mitsubishi Electric Bldg., Marunouchi, Chiyodaku, Tokyo, Japan.

San Ei Instrument Co. Ltd.,
1-89 Kashiwagi, Shinjuku-ku, Tokyo, Japan.

Fritz Schwartzer GMBH,
D-8000 München 60, Bärmannstrasse 38, Germany (BRD).

Smith & Nephew Research Ltd.,
Gilson Park, Harlow, Essex, U.K.

REFERENCES

References

1. Thermometers and Psychrometers

1963

PALLETT J.E. (1963) An electrical thermometer. *Electron. Engng.* **35, 313-315.**

1964

HEINING B. (1964) A thermistor wet and dry hygrometer probe. *J. agric. Engng Res.* **9, 96.**

SLATYER R.O. & BIERHUIZEN J.F. (1964) A differential psychrometer for continuous measurements of transpiration. *Pl. Physiol.* **39,** 1051-1056.

WALKER P.T. & TURNER M.L. (1964) A small temperature and humidity meter based on thermistors and humistors. *Rep. Trop. Pestic. Res. Unit, Porton.* **271.**

1965

FUCHS M. & TANNER C.B. (1965) Radiation shields for air temperature thermometers. *J. appl. Meteorol.* **4,** 544-547.

KOZAK B. (1965) Device for the recording of the soil temperature gradient. *Hivat. Kiad.* **28, 2,** 180-183. (in Hungarian)

SARGEANT D. H. (1965) Note on the use of junction diodes as temperature sensors. *J. appl. Meteorol.* **4,** 644-646.

VOGEL T.C. & JOHNSON P.L. (1965) Evaluation of an economical instrument shelter for micrometeorological studies. *Forest Sci.* **11, 4,** 434-435.

1967

HUMPHREY S.J.E. & WOLFF H.S. (1967) A temperature SAMJ. *J. Physiol.* **194,** 5-6.

KOZMA F. (1967) Measurement and recording of temperature by using thermistors. *Hivat. Kiad.* **33,** 163-185. (in Hungarian)

MAHER F.J. (1967) The multivibrator bridge for temperature measurement. *J. scient Instrum.* **44,** 531-534.

POLAVARAPU R.J. & MUNN R.E. (1967) Direct measurement of vapour pressure fluctuations and gradients. *J. appl. Meteorol.* **6,** 699-706.

SARGEANT D.H. & TANNER C.B. (1967) A simple psychrometric apparatus for Bowen ratio determinations. *J. appl. Meteorol.* **6**, 414-418.

STEVENS D.W. (1967) High-resolution measurement of air temperatures and temperature difference. *J. appl. Meteorol.* **6**, 179-185.

1968

ACOCK B. (1968) Methods of measuring leaf temperature. *Acta Hortic.* 7, 74-80.

FRAYER J.W. (1968) A method for constructing and installing thermocouples for measuring soil temperature. *Can. J. Soil Sci.* **48**, 366-368.

LONG I.F. (1968) Instruments and Techniques for Measuring the Microclimate of Crops. *In: The Measurement of Environmental Factors in Terrestrial Ecology. Symp. Brit. Ecol. Soc.* ed. R.M. Wadsworth, Blackwell Scientific Publications, Oxford.

MACFADYEN A. & WEBB N.R.C. (1968) An improved temperature indicator for use in ecology. *Oikos.* **19**, **1**, 19-27.

MAHER F.J. (1968) A sensitive thermometer. *J. scient. Instrum., Ser. 2.* **1**, 584-585.

SMIRNOV S.A. (1968) On the problem for the technique for measuring the temperature of the surface of the soil. *Gl. Geophys. Obs., Leningrad, T. Vyp.* **230**, 92-98. (in Russian)

1969

ALON Y. & JONAS M. (1969) Thermistor thermometer for linearized magnetic recording and telemetry. *Rev. scient. Instrum.* **40**, 646-647.

CHRISTOFFEL D.A. & CALHAEM I.M. (1969) A geothermal heat flow probe for 'in situ' measurement of both temperature gradient and thermal conductivity. *J. scient. Instrum., Ser. 2.* **2**, 457-465.

COTTON R.F. (1969) A small resistance thermometer psychrometer for use in high solar radiation conditions. *J. agric. Engng.* **14**, **3**, 284-289.

COTTON R.F. (1969) The droplet psychrometer for the measurement of aerial environment. *Agric. Meteorol.* **6**, **6**, 457-462.

HARTLEY G.S. & MACLAUCHLIN J.W.G. (1969) A simple integrating thermometer for field use. *J. Ecol.* **57**, 151-154.

LOURENCE F.J. & PRUITT W.O. (1969) A psychrometer system for micrometeorological profile determination. *J. appl. Meteorol.* **8**, 492-498.

RIELEY J.O., MACHIN D. & MORTON A. (1969) The measurement of microclimatic factors under a vegetation canopy. *J. Ecol.* **57**, **1**, 101-108.

ROSENBERG N.J., BROWN K.W. & SANDIN D.E. (1969) A thermocouple assembly for Bowen ratio and aerodynamic profile measurements of temperature and humidity. *Univ. Nebraska Hortic. Prog. Rep.* **73**, 28-44.

THULIN A. (1969) Composite thermocouple resistance sensors with linear output. *J. scient. Instrum., Ser. 2.* **2**, 1069-1072.

1970

ANDERSON A. (1970) A multi-channel temperature monitor. *J. scient. Instrum., Ser. 2.* **3**, 782-784.

HINSHAW R. & FRITSCHEN L.J. (1970) Diodes for temperature measurement. *Proc. 3rd. For. Microclimat. Symp. Canad. For. Ser.* 119.

LINACRE E.T. & HARRIS W.J. (1970) A thermistor leaf thermometer. *Plant Physiol.* **46**, 190-193.

PAINTER H.E. (1970) A recording resistance psychrometer. *Met. Mag., Lond.* **99**, **1172**, 68-75.

WEAVING G.S. (1970) An improved thermistor temperature integrator. *J. scient. Instrum., Ser. 2.* **3**, 711-714.

2. Hygrometers and Infra-Red Analysers

1956

TANNER C.B. & SUOMI V.E. (1956) Lithium chloride dewcel properties and use for dew-point and vapor-pressure gradient measurements. *Trans. Am. geophys. Un.* **37**, 413-420.

1958

TANNER C.B. & SUOMI V.E. (1958) A max-min dew-point hygrometer. *Trans. Am. geophys. Un.* **39**, 63-66.

1960

JONES F.E. & WEXLER A. (1960) A barium fluoride film hygrometer element. *J. geophys. Res.* **65**, 7, 2087-2095.

1961

BORGHORST A.J.W. (1961) Recording of relative humidity with semiconductor elements. *Arch. Met. Geophys. Bioklim. Ser. B.* **10**, 4.

1964

WENTZEL J.D. (1964) Humidity measurement. A critical survey. *CSIRO Spec. Rep. MEG* **269**.

1965

ACHESON D.T. (1965) The lithium bromide dew-cell. *J. appl. Meteorol.* **4**, 646-648.

BOUYOUCOS G.J. & COOK R.L. (1965) Humidity sensor; permanent electric hygrometer for continuous measurement of the relative humidity of the air. *Soil Sci.* **100**, 63-67.

1966

FALK S.O. (1966) A microwave hygrometer for measuring plant transpiration. *Z. Pflanzenphysiol.* **55**, **1**, 31-37. (in German)

1967

BACHEM C., JOHN G. & RUST G. (1967) A new instrument for measuring dew-points. *Met. Rdsch.* **20**, 67-68. (in German)

1968

BEGG J.E. & LAKE J.V. (1968) Carbon dioxide measurement; a continuous conductimetric method. *Agric. Meteorol.* **5**, **4**, 283-290.

LAI J.R. & HIDY G.M. (1968) Microsensor for measuring humidity. *Rev. scient. Instrum.* **39**, 1197-1203.

LEGG J.B. & PARKINSON K.J. (1968) Calibration of infra-red gas analyser for use with carbon dioxide. *J. scient. Instrum., Ser. 2.* **1**, 1003-1006.

1969

BOLLEN D. (1969) A thermistor hygrometer. *Wireless Wld., London.* **76**, 557-561.

BOWMAN G.E. (1969) Carbon dioxide assimilation measurement in a controlled environment glasshouse. *J. agric. Engng.* **14**, **1**, 92-99.

1970

ENOCH H. (1970) A new portable carbon dioxide gas analyser and its use in field measurements. *Agric. Meteorol.* **7**, **3**, 255-262.

3. Solarimeters (Pyranometers and Pyrheliometers)

1962

IMPENS I. (1962) Uses of the Bellani spherical pyranometer, type Davos. *Meded. Landb-Hogesch. OpzoekStns Gent.* **27**, **2**, 575-580.

MONTEITH J.L. & SZEICZ G. (1962) Simple devices for radiation measurement and integration. *Arch. Met. Geophys. Bioklim. Ser. B.* **11**, **4**, 491-500.

1963

VEZINA P.E. (1963) The field performance of ten Bellani radiation integrators. *For. Chron.* **39**, **4**, 401-402.

1964

BORREL F. & LAUMONIER R. (1964) Study of a silicon cell pyranometer. *Notes Etab. Etudes Rech., Met. Paris.* **183.** (in French)

MAJUMDAR N.C. & KAUSHIK S.B. (1964) Simple instruments for measurements of direct solar radiation. *Sol. Energy.* **8**, 91-94.

SEKIHARA K. (1964) The amount and properties of solar radiation in Japan and the instruments for its measurement. *Sol. Energy.* **1**, 456-466.

SZEICZ G., MONTEITH J.L. & DOS SANTOS J. (1964) A tube solarimeter to measure radiation among plants. *J. appl. Ecol.* **1**, **1**, 169-174.

1965

FEDERER C.A. & TANNER C.B. (1965) A simple integrating pyranometer for measuring daily solar radiation. *J. geophys. Res.* **76**, 2301-2306.

SZEICZ G. (1965) A miniature tube solarimeter. *J. appl. Ecol.* **2**, **1**, 145-147.

YAZAKI K. (1965) Solar cell type of sunshine recorder. *Weather Service Bull., Tokyo.* **32**, 215-217. (in Japanese)

1966

KYLE T.G. (1966) A crystal radiometer with frequency-modulated output. *J. scient. Instrum.* **43**, **10**, 750-753.

YAZAKI K. & KORNHER A. (1966) A solar-cell sunshine recorder. *J. met. Res., Tokyo.* **18**, 30-48. (in Japanese)

1967

ANDERSON M.C. (1967) The role of heat transfer in the design and performance of solarimeters. *J. appl. Meteorol.* **6**, 941-947.

BRACH E.J. & MACK E.R. (1967) A radiant energy meter and integrator for plant growth studies. *Can. J. Bot.* **45**, 2081-2085.

KERR J.P., THURTELL G.W. & TANNER C.B. (1967) An integrating pyranometer for climatological observer stations and mesoscale networks. *J. appl. Meteorol.* **6**, 688-694.

KYLE T.G. (1967) A crystal pyrradiometer. *Geofis. pura. appl.* **66**, 126-132.

1968

BRINGMAN M. & RODSKJER N. (1968) A small thermoelectric pyranometer for measurement of solar radiation in field crops. *Arch. Met. Geophys. Ser. B.* **16**, 4, 418-433.

DIRMHIRN I. (1968) On the use of silicon cells in meteorological studies. *J. appl. Meteorol.* 7, 702-707.

HOROWITZ J.L. (1968) An easily constructed shadow-band for separating direct and diffuse solar radiation. *Sol. Energy.* **12**, 543-545.

NORRIS D.J. & TRICKETT E.S. (1968) A simple low cost pyranometer. *Sol. Energy.* **12**, 251-253.

SUMNER C.J. (1968) A solarimeter stand with motorized occulting disc. *J. appl. Meteorol.* 7, 145-147.

SUMNER C.J. & PATTERSON G. (1968) A sun-tracking radiation instrument stand. *Sol. Energy.* **12**, 537-542.

UCHIJIMA Z. (1968) A newly devised solarimeter for measuring photosynthetically active radiation. *Japan Agricultural Research Quarterly.* 3, 3, 20-22.

1969

BYRNE G.F., ROSE C.W. & TORSSELL B.W.R. (1969) Resistive element solarimeter. *Agric. Meteorol.* **6**, **6**, 453-456.

NORMAN J.M. & TANNER C.B. (1969) Transient light measurements in plant canopies. *Agron. J.* **61**, 6, 847–849

TANNER C.B., FEDERER, C.A., BLACK T.A., & SHAW J.B. (1969) An economical radiometer, its theory, performance and construction. *Univ. Wis. Coll. Agric. Life Sci. Res. Rep.* **40**.

1970

GARDNER G. & SUTTON F. (1970) A portable linear amplifier for use with solarimeters. *J. appl. Ecol.* 7, 653-656.

PALTRIDGE G.W. (1970) A filter for absorbing photosynthetically-active radiation and examples of its use. *Agric. Meteorol.* 7, 167-174.

4. Net Radiometers

1960

TANNER C.B., BUSINGER J.A. & KUHN P.M. (1960) The economical net radiometer. *J. geophys. Res.* **65**, No. 11, 3657-3667.

1962

FUNK J.P. (1962) A net radiometer designed for optimum sensitivity and a ribbon thermopile used in a miniaturized version. *J. geophys. Res.* **67**, No. 7, 2753-2760.

TUNMORE B.G. (1962) A simple radiometer for the measurement of radiative heat exchange between buildings and the environment. *J. scient. Instrum* **39**, 219-221.

1963

FRITSCHEN L.J. (1963) Construction and evaluation of a miniature net radiometer. *J. appl. Meteorol.* **2**, 165-172.

1964

CAMPBELL G.S., ASHCROFT G.L. & TAYLOR S.A. (1964) Thermistor sensor for the miniature net radiometer. *J. appl. Meteorol.* **3**, 640-642.

1965

FRITSCHEN L.J. (1965) Miniature net radiometer improvements. *J. appl. Meteorol.* **4**, 528-542.

1966

COLLINS B.G. & KYLE T.G. (1966) The spectral variation of the sensitivity of a polythene shielded net radiometer. *Pure & appl. Geophys., Basel.* **63**, **1**, 231-236.

1967

PAULSEN H.S. (1967) Some experiences with the calibration of radiation balance meters. *Arch. Met. Geophys. Bioklim., Ser. B.* **15**, 156-174.

1969

BROWN K.W. (1969) Development of a sensor for measurement of net radiation within a crop canopy. *Univ. Nebraska Hortic. Progress Rep.* **73**, 45-55.

PALTRIDGE G.W. (1969) A net long-wave radiometer. *Q. Jl. R.met. Soc.* **95**, 635-638.

1970

COLLINS B.G. (1970) A laboratory system for short-wave calibration of net pyrradiometers. *Pure & appl. Geophys.* **80**, 371-377.

IDSO S.B. (1970) The relative sensitivity of polythene shielded net radiometers for short and long wave radiation. *Rev. scient. Instrum.* **41**, 939-943.

5. Other Radiation Instruments

1963

KUBIN S. & HIADEK L. (1963) An integrating recorder for photosynthetically active radiant energy with improved resolution. *Plant Cell Physiol., Prague.* **4**, **2**, 153-168. (in Czech)

McKEE G.B. (1963) A self-powered light integrator for ecological research. *Agron. J.* 55, 6, 580-583.

ROBERTSON G.W. & HOLMES R.M. (1963) A spectral light meter; its construction, calibration and use. *Ecology.* **44**, **2**, 419-423.

1964

ANDERSON M.C. (1964) Light relations of terrestrial plant communities and their measurement. *Biol. Rev., Cambridge Phil. Soc.* **39**, **4**, 425-486.

BORREL F. & LAUMENIER R. (1964) A pyranometer integrator with a photo-electric cell. *Notes Etab. Etudes Rech. Met., Paris.* **183**. (in French)

CALLAHAN P.S. (1964) An inexpensive actinometer for continuous recording of moonlight, daylight or low intensity evening light. *J. ecol. Entomol.* 57, **5**, 758-760.

DOWNS R.J. (1964) Measurement of irradiance for plant growth and development. *Proc. Am. Soc. Hort. Sci.* **85**, 663-671.

HUXLEY P.A. (1964) Performance of the Megatron-Siemens integrating photometer. *J. agric. Engng. Res.* **9**, **3**, 225-229.

LOGAN K.T. & PETERSON E.B. (1964) A method of measuring and describing light patterns beneath a forest canopy. *Publ. Dept. Forestry Can.* **1073**.

1965

COMBS A.C. (1965) Application of infra-red radiometers to meteorology. *J. appl. Meteorol.* **4**, 253-262.

McCREE K.J. (1965) Light measurements in plant growth investigations. *Nature, London.* **206**, **4983**, 527-528.

1966

FEDERER C.A. & TANNER C.B. (1966) Sensors for measuring light available for photosynthesis. *Ecology.* **47**, **4**, 654-657.

FUCHS M. & TANNER C.B. (1966) Infra-red thermometry of vegetation. *Agron. J.* **58**, **6**, 597-601.

McCREE K.J. (1966) A solarimeter for measuring photosynthetically active radiation. *Agric. Meteorol.* **3**, 353-366.

STAIR R. (1966) The measurement of solar radiation with principal emphasis on the ultra-violet component. *Air & Wat. Pollut.* **10**, 665-688.

STOUTJESDIJK P.H. (1966) On the measurement of the radiant temperature of vegetation surfaces and leaves. *Wentia.* **15**, 191-202.

1967

BOWERS S.A. & HAYDEN C.W. (1967) A simple portable reflectometer for field use. *Agron. J.* **59**, 490-492.

BRACH E. & WIGGINS B.W.E. (1967) A portable spectrophotometer for environmental studies of plants. *Lab. Pract.* **16**, 302-309.

JACKSON J.E. & SLATER C.H.W. (1967) An integrating photometer for outdoor use, particularly in trees. *J. appl. Ecol.* **4**, 421-424.

KOJIMA C. & KITADA K. (1967) A simple integrating light meter and its application to the continuous measurement of insolation in several stands. *J. Jap. For. Soc.* **49**, **2**, 69-72. (in Japanese)

LULL H.W. & REIGNER I.C. (1967) Radiation measurements by various instruments in the open and in the forest. *U.S. Dept. Agr. Forest Serv. Prod. Res. Dept.* **NE-84.**

McCREE K.J. & MORRIS R.A. (1967) A transmission meter for photo-synthetically active radiation. *J. agric. Engng. Res.* **12**, **3**, 246-248.

1968

BERRY R.E. & RANEY L.W. (1968) A recording photometer for biological studies. *Ecology.* **49**, **1**, 161-162.

CHASTON I. & WALKER P.J. (1968) A probe photometer. *Oikos.* **19**, **1**, 19-27.

GATES D.M. (1968) Sensing biological environments with a portable radiation thermometer. *Appl. Opt.* **7**, 1803-1809.

GETZ L.L. (1968) A method for measuring light intensity under dense vegetation. *Ecology.* **49**, **6**, 1168-1169.

HEY E.N. (1968) Small globe thermometers. *J. scient. Instrum., Ser. 2.* **1**, 955-957.

LIDWELL O.M. & WYON D.P. (1968) A rapid response photometer for out-door use, particularly in trees. *J. appl. Ecol.* **4**, 421-424.

McCREE K.J. (1968) Infra-red sensitive colour film for spectral measurements under plant canopies. *Agric. Meteorol.* **5**, **3**, 203-208.

SZEICZ G. (1968) Measurement of Radiant Energy. *In: Measurement of Environmental Factors in Terrestrial Ecology. Symp. Brit. Ecol. Soc.* ed. R.M. Wadsworth, Blackwell Scientific Publications, Oxford.

UCHIJIMA Z. (1968) A newly devised solarimeter for measuring photo-synthetically active radiation. *Jap. agric. Res.* **3, 3,** 20-22.

VYNCKE G. & GOOSENS R. (1968) An apparatus for the instantaneous measurement of the relative light intensity. *Sylva gandavensis, Gent.* **11**.

1969

DAYNARD T.B. (1969) An inexpensive space-integrating light meter for field crop research. *Can. J. Pl. Sci.* **49, 2,** 231-234.

IDSO S.B. & JACKSON R.D. (1969) A method for the determination of the infra-red emittance of leaves. *Ecology.* **50, 5,** 899-902.

McPHERSON H.G. (1969) Photocell filter combinations for measuring photo-synthetically active radiation. *Agric. Meteorol.* **6, 5,** 347-356.

MATTSSON J.D. (1969) Infra-red photography; a new technique in micro-climate investigations. *Weather, London.* **24, 3,** 107-112.

NORMAN J., TANNER C.B. & THURTELL G.W. (1969) Photosynthetic light sensor for measurements in plant canopies. *Agron. J.* **61, 6,** 840-843.

1970

KENDALL J.M. & BERDAHL C.M. (1970) Two blackbody radiometers of high accuracy. *Appl. Opt.* **9,** 1082-1091.

KUEHN L.A., STUBBS R.A. & WEAVER R.S. (1970) Theory of the globe thermometer. *J. appl. Physiol.* **29,** 750-757.

MATTSSON J.O. (1970) Infra-red thermography – a new technique in microclimatic investigations. *Weather.* **24,** 107-112.

STRONG J. & HOPKINS G.W. (1970) An environmental mean radiation temperature meter. *Rev. scient. Instrum.* **41,** 360-364.

6. Heat Flow Meters

1961

PHILIP J.R. (1961) The theory of heat flux meters. *J. geophys. Res.* **66,** 571-579.

1965

HAGER N.E. (1965) Thin foil heat meter. *Rev. scient. Instrum.* **36,** 1564-1570.

1968

FUCHS M. & TANNER C.B. (1968) Calibration and field test of soil heat flux plates. *Soil Sci. Soc. Am. Proc.* **32,** 326-328.

1969

POPPENDIEK H.E. (1969) Why not measure heat flux directly? *Env. Quarterly.* **15,** No. 1.

1970

MORGENSEN V.O. (1970) The calibration factor of heat flux meters in relation to the thermal conductivity of the surrounding medium. *Agr. Meteorol.* **7,** 401-410.

7. Anemometers

1959

LEWIS T. & SIDDORN J.W. (1959) A simple portable recorder for direction and speed of wind. *J. Anim. Ecol.* **28**, 377-380.

1964

GIGLIOLI M.E.C. (1964) A simple recording wind gauge. *Mosquito News.* **24**, 377-382.

HOLMES R.M., GILL G.C. & CARSON H.W. (1964) A propeller-type vertical anemometer. *J. appl. Meteorol.* **3**, **6**, 802-804.

MOUNT L.E. (1964) A simple instrument for the measurement of low air speeds, with special reference to pig housing. *J. agric. Sci.* **63**, 335-339.

1965

HEAD M.R. & SURREY N.B. (1965) Low speed anemometer. *J. scient. Instrum.* **42**, 349.

SAYER H.J. (1965) A remote reading anemometer using an electric tachometer. *Weather, London.* **20**, 383-386.

SUMNER C. J. (1965) A long period recorder for wind speed and direction. *Q. J. Roy. Met. Soc.* **91**, 346-347.

1966

HOFMANN G. (1966) Measurement of low wind velocities with cup anemometers. *Z. Meteorol.* **17**, 335-338. (in German)

MORRIS R.A., MORMAN A.R. & SCOTT D. (1966) A portable integrating omnidirectional anemometer. *Radio & Electron. Eng.* **32,** No. 6, 371-376.

1967

BERNSTEIN A.B. (1967) A note on the use of cup anomemeters in wind profile experiments. *J. appl. Meteorol.* **6**, 280-286.

FRITSCHEN L.J. (1967) A sensitive cup-type anemometer. J. *appl. Meteorol. 6*, 695-698.

HARTLEY G.E.W. (1967) A cup anemometer suitable for construction in school metal workshops. *Weather, London.* **22,** 416-418.

WYNGAARD J.C. & LUMLEY J.L. (1967) A constant temperature hot-wire anemometer. *J. scient. Instrum.* **44**, 363-365.

1968

LONG I. F. (1968) Instruments and Techniques for Measuring Microclimate of Crops. In: *The Measurement of Environmental Factors in Terrestrial Ecology. Symp. Brit. Ecol. Soc.* ed. R. M. Wadsworth, Blackwell Scientific Publications, Oxford.

MARSHALL J.K. (1968) An automatic battery operated wind-speed and direction recording system. *Nature & Resour., Unesco.* 7, 309-311.

SUMNER C.J. (1968) A low torque cup anemometer. *Aust. J. Inst. Control.* **Oct. 1968**, 39-40.

1969

BRADLEY E.F. (1969) A small sensitive anemometer system for agricultural meteorology. *Agric. Meteorol.* **6**, **3**, 185-194.

BYASS J.B. & RANDALL J.M. (1969) Equipment and methods for orchard spray application research. IV. A directional anemometer for measuring airflow in orchard canopies. *J. agric. Engng.* **14**, **2**, 187-194.

KOVASZNAY L.S.G. & CHEVRAY R. (1969) Temperature compensated linearizer for a hot-wire anemometer. *Rev. scient. Instrum.* **40**, 91-94.

1970

DIBLEY G.C. & LEWIS T. (1970) A bivane direction indicator and a sensitive vertical anemometer for measuring components of wind in sheltered places. *Ann. appl. Biol.* **66**, 469-475.

JONES J.I.P. (1970) A new recording wind-vane. *J. scient. Instrum., Ser. 2.* **3**, 9-14.

MEYER G. & STEINER A. (1970) On measurements with a hot-wire anemometer at constant temperature. *C. R. Acad. Sci., Paris. 270 Ser. A.* 155-158. (in French)

SADEH W.Z., MAEDER P.F. & SUTERA S.P. (1970) A hot-wire method for low velocity with large fluctuations. *Rev. scient. Instrum.* **41**, 1295-1298.

ZELDIN B. & SCHMIDT F.W. (1970) Calibrating a list film anemometer for low velocity measurement in non-isothermal flow. *Rev. scient. Instrum.* **41**, 1373-1374.

8. Water Content Meters

(Neutron Soil Water Meters, Surface Wetness Meters etc.)

1960

JONES R.L. (1960) A simple dewfall integrator. *N.Z. J. Sci.* 7, 45-50.

1962

CANARACHE A. (1962) Determination of soil moisture content by a thermostatic method. *Studii Cerc. Agron. Cluj.* **13**, 113-123. (in Rumanian)

JOHNSON A.I. (1962) Methods of measuring soil moisture in the field. *U.S. Geol. Surv., Water Supply Papers.* **1619U**.

RICHARDS L.A. (1962) Tests of instruments for measuring water status in field soils. *Agr. Res. Serv. Salinity Lab. Ann. Rep.* 1-5.

1963

BEARD J.S. (1963) Soil moisture studies using the Berkeley fibreglass instrument. *S. Afr. Forest J.* **44**, 24-28.

SHALIBASHVILI A. (1963) Comparative evaluation of methods of determining soil moisture. *Trudy pochv. Inst.* **11**, 243-248. (in Russian)

1964

DEL VALLE FLORENCIA F. (1964) Determination of soil moisture with gypsum blocks and the 'Chapingo 63' recorder. *Ingr. Hidrol. Mex.* **18**, 46-49. (in Spanish)

FUJIOKO Y. & NISHIDE T. (1964) Glass-filter electrical moisture block and self-recording moisture meter for measuring soil moisture. *Trans. agric. Engng. Soc., Tokyo.* **9**, 5-10. (in Japanese)

LINDNER H. (1964) An instrument for measuring the water content and temperature of soil by means of thermistors. *Albrecht-Thaer-Arch.* **8**, 79-87. (in German)

MYHR E. (1964) A new tensiometer. *Grundforbattring.* **17**, 181-184. (in Swedish)

PRZESMYCKI J. (1964) Determination of soil moisture by the Bouyoucos method. *Gaz. Obs. Panst. Inst. Hydrol. Met., Warsaw.* **17**, **5**, 5-7. (in Polish)

RIZZO A.F. (1964) Studies in moisture and temperature in the La Plata soil profiles. Tensiometer techniques (Richards Method). *Revta. Fac. Agron. Univ. nac. La Plata.* **40**, 219-256. (in Spanish)

VAN DER WESTHUIZEN M. (1964) On the possibility of measuring soil moisture with high frequency electromagnetic waves. *S. Afr. J. agric. Sci.* 7, 589-590.

1965

AKOPOV R.M. (1965) A dielectric soil moisture meter. *Vest. sel'. – khoz. Nauki, Mosk.* **11**, 130-131. (in Russian)

BABALOCA J.M. (1965) Soil moisture measurement. *Proc. Symp. Trop. Met., Oshodi, Lagos.* **Aug. 1964**: 90-95.

CAPIEL M. (1965) Evaluation of the intrinsic performance of electrical resistance units for measuring soil moisture. *J. Agr. Univ. P. Rico.* **49**, 350-367.

COPE F. & TRICKETT E.S. (1965) Measuring soil moisture. *Soil & Fertilizers.* **28**, **3**, 201-206.

DANILIN A.I. (1965) Photoelectric method of determining the moisture content of soils. *Pochvovedenie.* **11**, 90-95. (in Russian)

LOMAS J. (1965) Note on dew-duration recorders under semi-arid conditions. *Agric. Meteorol.* **2**, **5**, 351-360.

McILROY I.C. (1965) A compact tensiometer with an inbuilt manometer. *J. agric. Engng. Res.* **10**, **2**, 183-187.

NEGOVELOV S.F. & ZHULIP L.P. (1965) Determination of soil moisture available to plants under field conditions. *Pochvovedenie.* **9**, 81-86. (in Russian)

NEWBOULD P., MERCER E.R. & LAY P.M. (1965) Change in water content at different depths of soil; a new method of measurement. *Agr. Res. Council Radiobiol. Lab. Rep.* **1964-65**, 51-55.

NOFFSINGER T.L. (1965) Survey of techniques for measuring dew. *Humidity and Moisture*, ed. Wexler, Rheinhold, New York. **2**, 523-531.

SAINI G.R. (1965) Principle of the immersion method for soil moisture measurement. *Soil Sci.* **100**, 220.

VAN BAVEL C.H.M. (1965) Neutron scattering measurement of soil moisture; development and current status. *Humidity and Moisture*, ed. Wexler, Rheinhold, New York. **4**, 171-184.

ZALIVADNYI B.S. (1965) Inadequacy of the method of calibrating sorption moisture meters. *Pochvovedenie.* **3**, 100-102. (in Russian)

1966

BELL J.P. & McCULLOCH J.S.G. (1966) Soil moisture estimation by the neutron scattering method in Britain. *J. Hydrol.* **4**, 254-263.

HUGHES M.W. (1966) The examination of the new electrical resistance block for measuring soil water suction. *J. Aust. Inst. agric. Sci.* **32**, 225-227.

LARSEN S. & WIDDOWSON A.E. (1966) Determination of soil moisture by electrical conductivity. *Soil Sci.* **101**, 420.

MACK A.R. & BRACH E.J. (1966) Soil moisture measurement with ultra-sonic energy. *Proc. Soil Sci. Soc. Am.* **30**, **5**, 544-548.

PECK A.J. & RABBIDGE R.M. (1966) Soil water potential; direct measurement by a new technique. *Science.* **151**, **3716**, 1385-1386.

STEVENS G.N. & HUGHES M. (1966) Moisture meter performance in field and laboratory. *J. agric. Res.* **11**, **3**, 210-217.

THOMAS A.M. (1966) On site measurements of moisture in soil and similar substances by 'fringe capacitance'. *J. scient. Instrum.* **43**, 21-27.

WEBSTER R. (1966) The measurement of soil moisture tension in the field. *New Phytol.* **65**, 249-258.

ZALIVADNYI B.S. (1966) Inadequacy of the method of calibrating sorption moisture meters. *Pochvovedenie.* **3**, 100-102. (in Russian)

1967

BOODT M. de. (1967) Neutron method of soil moisture determination. *Proc. int. Soil Water Symp., Praha.* 109-128.

DOBRZANSKI B. (1967) New electrical resistance method for continuous measurements of soil moisture content under field conditions. *Proc. int. Soil Water Symp., Praha.* 179-187.

HOLMES J.W., TAYLOR S.A. & RICHARDS S.J. (1967) Measurement of soil water. *Agronomy.* **11**, 273-303.

HUNGERFORD K.E. (1967) An acetate dew gage. *J. appl. Meteorol.* **6**, 936-940.

KASPAR I. (1967) Some remarks on soil moisture measurement by a thermal method. *Proc. int. Soil Water Symp., Praha.* 167-177.

PAPP B. (1967) Measurement of soil moisture. *Hivat. Kiadv.* **33**, 186-217. (in Hungarian)

SZUNIEWICZ J. (1967) A tensiometer with a mercury manometer for determining the soil moisture potential under field conditions. *Wiadonosci Inst. Melior.* 7, **1**, 105-125. (in Polish)

VAN BAVEL C.H.M. & STIRK G.B. (1967) Soil water measurement with an Am^{241}-Be neutron source and an application to evaporimetry. *J. Hydrol.* **5**, 40-46.

1968

BALLY R.J. (1968) About pyrometric determination of soil moisture. *Met. Gidrol. Gospod. Apelor.* **13**, 325-331. (in Rumanian)

CZUBEK J.A. (1968) Range of nuclear methods for soil moisture and density measurements. *Nukleonika.* **13**, 517-533. (in Polish)

DIMAKSJAN J. & GLAZKOV N.P. (1968) The measurement of soil moisture by means of the cadmium ratio. *Met. i. Gidrol., Leningrad.* **9**, 68-74. (in Russian)

MYHILL R.R. (1968) Tensiometers for the soil-water research programme at Samaru. *Samaru Misc. Pap.* **24**.

PRIHAR S.S. & SANCHU B.S. (1968) A rapid method of soil moisture determination. *Soil Sci.* **105**, 142-144.

RIVOIRA G. & TEDEXHI P. (1968) *In situ* moisture determination by means of a neutron moisture meter. *Studi sassar.* **16 Sect. III.**, 578-593. (in Italian)

ZALIVADNYI B.S. (1968) Use of a dynamic heat-capacity moisture meter in soil science. *Pochvovedenie.* **2**, 149-158. (in Russian)

1969

BELL J.P. (1969) A new design principle for neutron soil moisture gauges; the Wallingford neutron probe. *Soil Sci.* **108**, 160-164.

BELL J.P. & McCULLOCH J.S.G. (1969) Soil moisture estimation by the neutron method in Britain. A further report. *J. Hydrol.* 7, 415-433.

BROWN K.W. & SANDIN D.E. (1969) Surface moisture duration detector. *Univ. Nebraska Hortic. Progress Rep.* **73**, 11-13.

CRISTEA N., PLACINTA A. & VASILESCO G. (1969) Some considerations of the methods for measuring dew. *Cul. de Lucrari, 1967, Inst. Met. Bucharest.* 423-439. (in Rumanian)

DE JAGER J.M. & CHARLES-EDWARDES J. (1969) Thermal conductivity probes for soil moisture determinations. *J. exp. Bot.* **20**, 46-51.

DENDAS J. (1969) The use of tensiometer and electrical measurements for the *in situ* determining of changes in moisture tension. *Pedologie, Gand.* **19**, 23-33. (in French)

DIMAKSJAN A.M. & GLAZKOV N.P. (1969) Radiometric methods of measuring soil moisture. *Gosudarst. Gidr. Inst., Leningrad. T.Vyp.* **168**, 3-30. (in Russian)

FILIPPOV L.A. (1969) Refractometric method of estimating the water regime of the tea plant. *Subtrop. Kul'tury.* **1**, 58-64. (in Russian)

LEDNICKY V. & RICHTER V. (1969) A contribution to soil moisture measurement. *Met. Zpravy, Prague.* **22**, 28-31. (in Czech)

RAWITZ E. (1969) Installation and field calibration of neutron-scattering equipment for hydrologic research in heterogenous and stony soils. *Wat. Resour. Res.* **5**, 519-523.

RIJTEMA P.E. (1969) The calculation of non-parallelism of gamma access tubes, using soil sampling data. *J. Hydrol.* 9, 206-212.

RYHINER A.H. & PANKOW J. (1969) Soil moisture measurement by the gamma transmission method. *J. Hydrol.* 9, 194-205.

SARTZ R.S. (1969) Interpreting nuetron probe readings in frozen soil. *U.S. For. Serv. Res.* **Note NC**-77, 4.

TOROK I. (1969) Determination of soil moisture with a neutron scattering apparatus. *Agrokem. Talajt.* **18**, 107-116. (in Hungarian)

VANDOVA M. (1969) On the accuracy of measurement of moisture in some soil varieties. *Hidr. Met., Sofia.* **18**, **4**, 43-50. (in Bulgarian)

1970

LOMAS J. & SHASHOUA Y. (1970) The performance of three types of leaf wetness recorders. *Agric. Meteorol.* 7, 2, 159-166.

McHENRY J.R. & GILL A.C. (1970) Measurement of soil moisture with a portable gamma-ray scintillation spectrometer. *Water Resour. Res.* **6**, 989-992.

STREBEL O. (1970) A pressure transducer-tensiometer for automatic recording of soil moisture suction in the field. *Zeit. Pflanzenerah. Bodenk., Leipzig.* **126**, 6-15. (in German)

9. Water Vapour Flux Meters

1961

CUSTANCE A.C. & LAFLAMME C.R. (1961) A new technique for the continuous measurement and automatic recording of sweating rates. *Defence Research Chemical Laboratories Report* **No. 359**.

1963

LHOTA O. (1963) A new method of simultaneous measurement of evaporation from soil and transpiration by plants. *Rost. vyroba, Praha.* **36**, 1259-1274. (in Czech)

SKAU C.M. & SWANSON R.H. (1963) An improved heat pulse velocity meter as an indicator of sap speed and transpiration. *J. geophys. Res.* **68**, **16**, 4743-4749.

TEPE W. & LEIDENFROST E. (1963) A method of measuring the water supply to plants. *Z. Acker -u Pflbau.* **117**, 395-402. (in German)

WALKER D.C. (1963) New instrument for measuring evaporation; its application to the evaporation of water through thin films. *Rev. scient. Instrum.* **34**, 1006-1009.

1965

DYER A.J. & MAHER F.J. (1965) The 'evapotron'; an instrument for the measurement of eddy fluxes in the lower atmosphere. *CSIRO Div. Met. Phys., Tech. Pap.* **15**.

DYER A.J. & MAHER F.J. (1965) Automatic eddy-flux measurement with the evapotron. *J. appl. Meteorol.* **4**, **5**, 622-625.
KOZMA F. (1965) New methods for measuring transpiration. *Hivat. Kiadv.* **28**, **2**, 93-98 (in Hungarian)
LANGE O.L. (1965) Modern methods for measuring photosynthesis and transpiration of plants in the field. *Schriftenteche Forstl. Fak. Univ. Gottingen.* **33**, 91-102. (in German)
SMITH A.D., MIKKELSEN T.T. & WALKER P.H. (1965) Comparison of apparatus for the weekly measurements of evaporation. *Can. J. Pl. Sci.* **45**, **5**, 511-513.

1966

BAY R.R. (1966) Evaluation of an evaporimeter for peat bogs. *Wat. Resour. Res.* **2**, 437-442.
ORMROD D.P. & WOOLLEY C.J. (1966) Apparatus for environmental physiology studies. *Can. J. Pl. Sci.* **46**, **5**, 573-575.

1967

DYER A.J., HICKS B.B. & KING K.M. (1967) The fluxatron – a revised approach to the measurement of eddy fluxes in the lower atmosphere. *J. appl. Meteorol.* **6**, 408-413.
ERDOS L. (1967) Main sources of error in the measurement of potential evaporation. *Idoj. Magy. Met. Tars. Foly.* **71**, 10-22. (in Hungarian)
McILROY I.C. (1967) An energy partition evaporation recorder. *Aust. J. Instrum. Technol.* **23**, 120-122.
SHIMSHI D. (1967) The use of a field porometer for the study of water stress in plants. *Israel J. Agr. Res.* **14**, **4**, 137-144.
SHAN LEI YU & BRUTSAERT W. (1967) Evaporation from very shallow pans. *J. appl. Meteorol.* **6**, 265-271.
TANNER C.B. (1967) Measurement of evapotranspiration. *Agronomy.* **11**, 534-574.

1969

KANEMASU E.T., THURTELL G.W. & TANNER C.B. (1969) Design, calibration and field use of a stomatal diffusion porometer. *Pl. Physiol. Lancaster.* **44**, 881-885.
STANHILL G. (1969) A simple instrument for the field measurement of turbulent diffusion flux. *J. appl. Meteorol.* **8**, 509-513.
TURNER N.C., PEDERSON F.C. & WRIGHT W.H. (1969) An aspirated diffusion porometer for field use. *Conn. Agr. Exp. Sta., Spec. Bull. Soils.* **29**, **100**.
WILLIAMS C.N. & SINCLAIR R. (1969) A sensitive porometer for field use. *J. exp. Bot.* **29**, **63**, 81-83.

1970

BYRNE G.F., ROSE C.W. & SLATYER R.O. (1970) An aspirated diffusion porometer. *Agric. Meteorol.* **7**, **1**, 39-44.
HANAFISA T. (1970) A simple method for the measurement of water vapour flux. *J. met. Soc. Japan.* **48**, 259-262.
LEYTON L. (1970) Problems and techniques in measuring transpiration from trees. *Physiology of Tree Crops*, ed. Luckwill and Cutting. Academic Press, London.
MEIDNER H. (1970) A critical study of sensor element diffusion porometers. *J. exp. Bot.* **21**.
MONTEITH J.L. & BULL T.A. (1970) A diffusive resistance porometer for field use. II Theory, calibration and performance. *J. appl. Ecol.* **7**, 623-638.
STILES W. (1970) A diffusive resistance porometer for field use. I Construction. *J. appl. Ecol.* **7**, 617-622.

10. General Purpose Recorders

1963

FRITSCHEN L.J. & VAN BAVEL C.H.M. (1963) Micrometeorological data handling system. *J. appl. Meteorol.* **2**, 151-155.

1965

BRADLEY E.F. & WALL B.H. (1965) Digitization and integration of low-level fluctuating signals. *Rev. scient. Instrum.* **36**, 691-693.

CAIRNS E.G. & TEVEBAUGH A.D. (1965) Sensitive voltage measurement and recording system. *Rev. scient. Instrum.* **36**, **12**, 1726-1727.

ERICSON B. (1965) A photographic data recording system. Automatic collection of climatic data in a field experiment. *Inst. Skogsprod. Skogshogeskol. Stockholm, Rappt. Uppsatser.* **8.** (in Swedish)

VOISEY P.W., MACDONALD D.C. & HANSEN H. (1965) The performance of a portable temperature recorder for field experiments. *Can. J. Pl. Sci.*

1966

KAIMAL J.C., HAUGEN D.A. & NEWMAN J.I. (1966) A computer controlled micrometeorological observation system. *J. appl. Meteorol.* **5**, 411-420.

1967

SACKETT S.S. & DECOSTE J.H. (1967) A new mobile forest laboratory. *Fire Control Notes.* **28**, **4**, 7-9.

1968

BLACKWELL M.J. & BLACKBURN M.R. (1968) Crop Environment Data Acquisition. *In: The Measurement of Environmental Factors in Terrestrial Ecology. Symp. Brit. Ecol. Soc.* ed. R. M. Wadsworth, Blackwell Scientific Publications, Oxford.

CERNESCA A. (1968) The use of automatic data recording systems for climate-ecological studies within productivity research. *Photosynthetica.* **2**, **4**, 238-244. (in German)

RUTTER N. (1968) Instrumentation for a transect study of microclimate with special reference to the value of time-lapse photography in agricultural meterology. *Int. J. Biomet.* **12**, 3-9.

SUMNER C.J. (1968) Digital recording mechanism for linear displacements. *J. scient. Instrum., Ser. 2.* **1**, 1237-1239.

VANSTONE F.H. (1968) The development of a system of instrumentation for recording crop environments. *Rep. Welsh Pl. Breed. Stn.* **1967**, 135-160.

1969

CERNESCA A. (1969) Photographic data recording for use in mobile microclimatic stations. *Col. Ges. Forstw.* **86**, **1**, 49-58. (in German)

FOWLER W.B. (1969) A digital temperature monitor for photorecording. *U.S. For. Serv. Res.* **Note PNW-95.**

INOUE E., UCHIJIMA Z., SAITO T., ISOBE S. & UEMURA K. (1969) The 'Assimitron', a newly devised instrument for measuring CO_2 flux in the surface air layer. *J. agric. Met., Tokyo.* **25**, **3**, 165-172.

PATRICHI F. & CRISTEA N. (1969) Recording the input data necessary to compute items of the thermal balance. *Culegero de Lucrari, 1967, Inst. Met. Bucharest,* 381-390. (in Rumanian)

1970

ALLEN L.H. (1970) An operational system for sampling and sensing micrometeorological elements and for logging and processing micrometeorological data. *Proc. 3rd. Forest Microclimat. Symp., Can. For. Service.* 91-116.

SUTTON F. & RORISON I.H. (1970) The modification of a data logger for the recording of temperatures in the field using thermistor sensors. *J. appl. Ecol.* 7, 321-329.

11. Integrators

1963

TANNER C.B., THURTELL G.W. & SWAN J.B. (1963) Integration systems using a commercial coulometer. *Soil Sci. Amer. Proc.* 27, 478-481.

1964

HANKS R.J. & GARDNER H.R. (1964) Portable integrator for net radiation, total radiation and soil heat flux. *Soil Sci. Soc. Am. Proc.* **28**, 449-450.

THURTELL G.W. & TANNER C.B. (1964) Electronic integrator for micrometeorological data. *J. appl. Met.* **3**, 198-202.

1965

FUNK J.P. & DOWE D.G. (1965) An integrator for multichannel potentiometric recorders. *J. scient. Instrum.* **42**, **8**, 615-617.

WHILLIER A. & TOUT D. (1965) A new integrating instrument for measuring daily values of total solar radiation. *Sol. Energy.* **9**, 208-212.

1966

HILL R.H. (1966) An attachment for the Sumner long-term recorder that integrates a radiometer output. *J. scient. Instrum.* **43**, **11**, 829-830.

1967

BRACH E.J. & MACK A.R. (1967) A radiant energy meter and integrator for plant growth studies. *Can. J. Bot.* **45**, 2081-2085.

1968

McGRUDDY P.J. & HOPKINSON J.M. (1968) An inexpensive integrator for use with solarimeter. *Field Sta. Res. Div. Plant Ind., CSIRO.* 7, **1**, 13-20.

1969

CAIN J.C. (1969) A portable economical instrument for measuring light and temperature intensity-time integrals. *Hort Science.* **4**, 123-125.

GOLDWATER F. (1969) A simplified integrator for a net radiometer. *Sol. Energy.* **12**, 117-119.

HILL R.H. (1969) A direct current millivolt integrator. *J. scient. Instrum., Ser. 2.* **2**, 616-618.

HUGHES M.K. & LINCOLN E. (1969) A simple integrator for use with solarimeters. *Oikos,* **20,** 1, 161-165.

KUHN L. & PODOGROCKI J. (1969) New integrators in actinometry. *Panst. Inst. Hydr. Met., Warsaw Gazeta, Obs.* **22**, **10**, 14-16. (in Polish)

BIBLIOGRAPHY

Bibliography

Temperature Measurement

1. HERZFELD C.M. ed. (1963) *Temperature; its measurement and control in science and industry.*

Vol. 3:

Part 1. Basic concepts, standards and methods;

Part 2. Applied methods and instruments;
(Chapters on thermocouples and radiometers)

Part 3. Biology and medicine;
(Chapters on measurement of temperature in animals and man)

Rheinhold Publishing Corporation, New York.

2. SESTAK Z., CATSKY J. & JARVIS P.G. ed. (1971) *Plant photosynthetic production: manual of methods.*

Chapter 17 is a very comprehensive review of leaf temperature measurements.

W. Junk, The Hague.

Humidity Measurement

1. WEXLER A. ed. (1965) *Humidity and moisture. Measurement and control in science and industry.*

Vol. 1. Principles and methods of measuring humidity in gases;
(Sections on psychrometry, dew-point hygrometry, electrical hygrometry and infra-red analysers)

Vol. 2. Applications;
(The measurement of humidity in meteorology, biology, medicine, agriculture and ventilation engineering)

Vol. 3. Fundamentals and standards;
(Properties of humid air; calibration and testing of humidity instruments)

Vol. 4. Principles and methods of measuring moisture in liquids and solids. (Chemical, electrical and nuclear methods)

Rheinhold Publishing Corporation, New York.

2. SPENCER-GREGORY H. & ROURKE E. (1957) *Hygrometry.*

Deals with the theory and behaviour of a wide range of hygrometers.

Crosby Lockwood & Son, London.

Gas Analysis

1. HILL D.W. & POWELL T. (1968) *Non-dispersive infra-red gas analysis in science, medicine and industry.*

The theory, performance and calibration of infra-red analysers.

Hilger Ltd., London.

2. SESTAK Z., CATSKY J. & JARVIS P.G. ed. (1971) *Plant photosynthetic production: manual of methods.*

Chapters 3 and 4 deal with infra-red and other CO_2 analysers in great detail; Chapter 5 with measurement of CO_2 exchange in the field.

W. Junk, The Hague.

Radiation Measurement

1. ROBINSON N. ed. (1966) *Solar radiation.*

Chapter 7 is concerned with radiation instruments.

Elsevier, Amsterdam.

2. SELLERS W.D. (1965) *Physical climatology.*

Chapter 6 describes radiation instruments.

University of Chicago Press, Chicago.

3. IGY INSTRUCTION MANUAL, PART VI. (1957) *Radiation instruments and measurements.*

Information about standard methods of calibrating and operating radiation instruments.

Pergamon Press, London.

4. ANON. (1966) *Meteorological Researches No. 15.*

Eleven articles describing the calibration and performance of Russian radiation instruments (Russian and English abstracts).

Nauka, Moscow.

5. ANON. (1967) *Methods of measuring photosynthetically active radiation (Russian).*
Nauka, Moscow.

Anemometers

OWER E. & PANKHURST R.C. (1966) *The measurement of air flow.*

Mainly concerned with pressure tube devices but with sections on hot-wire and vane anemometers.

Pergamon Press, Oxford.

Wetness Meters

W.M.O. TECHNICAL NOTE No. 55. (1963) *The influence of weather conditions on the occurrence of apple scab.*

The Appendix of this bulletin – longer than the main text – consists of full descriptions of 12 commercial and laboratory instruments for measuring the wetness of surfaces exposed to rain and dew.

W.M.O. Geneva.

Human Environments

BEDFORD T. (1964) *Basic principles of ventilation and heating. 2nd Edition.*

Includes a discussion of the principles of the katathermometer and globe thermometer.

H.K. Lewis, London.

Measuring and Recording Systems

1. DOEBELIN E. (1966) *Measurement systems: application and design.*

Intended mainly as a laboratory manual but containing several sections relevant to the measurement and recording of temperature, radiation, wind etc.

McGraw Hill Co., New York.

2. BRADLEY E.F. & DENMEAD O.T. ed. (1967) *The collection and processing of field data.*

Proceedings of a symposium: about half the book deals with general principles of collecting and analysing field measurements. Several chapters deal specifically with micrometeorological measurements.

Interscience, New York.

3. ENGINEERING EQUIPMENT USERS ASSOCIATION HANDBOOK No. 28 (1968)

Specification and selection of data logging equipment.

Constable, London.

4 CHUDNOWSKI A.F. & SLIMOVITCH B.M. (1966) *Semiconductors, radio-electronics and cybernetics in agrometeorology (in Russian).*

Hydrometeorology Publishing House, Leningrad.

IBP Handbooks

Several IBP Handbooks contain references to meteorological and micrometeorological instruments and methods.

VOLLENWEIDER R.A. ed. (1969) *Primary production in aquatic environments (IBP No. 12)*

This book has a complete section on the measurement of light under water (pp. 158-171).

Blackwell Scientific Publications, Oxford.

Brief statements on the measurement of temperature, humidity, wind and radiation, and on recording systems will be found in the following:

1. GOLLEY F.B. & BUESHNER H.K. ed. (1969) *Productivity of large herbivores (IBP No. 7)*. 146-151.
2. WEINER J.S. & LOURIE J.A. ed. (1969) *Human biology (IBP No. 9)*. 600-603.
3. PETRUSEWICZ K. & MACFADYEN A. (1970) *Productivity of terrestrial animals (IBP No. 13)*. 148-149.

Blackwell Scientific Publications, Oxford.

General

1. WADSWORTH R.M. ed. (1968) *The measurement of environmental factors in terrestrial ecology.* British Ecological Society Symposium No. 8.

   Chapters on measurement of temperature, humidity, wind, radiation, CO_2, soil water, soil aeration, data capture and telemetry.

   Blackwell Scientific Publications, Oxford

2. ECKARDT F.E. ed. (1965) *Methodology of plant eco-physiology.* Arid zone research XXV, Proc. Montpellier Symposium.

   Papers on measurement of radiation, energy balance, evaporation, soil and plant water potential, rainfall interception, carbon dioxide exchange, stomatal resistance.

   UNESCO, Paris.

3. ANON. (1966) *Conference for instrumentation on plant environment.*

   A number of short papers, several of which describe micrometeorological measurements and instruments.

   Soc. Instr. Tech., Australia.

4. MUNN R.E. (1971) *Biometeorological methods.*

   Chapters on the design of biometeorological experiments and statistical analysis.

   Academic Press, New York.

5. SLATYER R.O. & McILROY I.C. (1961) *Practical microclimatology.*

This book has inexplicably remained out of print since the rapid sale of the first edition but is still a useful introduction to methods of measuring rain, dew, evaporation and soil water.

UNESCO, Paris

ANON. (1970) *Techniques d'étude des facteurs physique de la biosphere.*

Thirty-nine authors from different centres in France contributed chapters describing the construction and performance of a wide range of instruments for measuring microclimate and plant water relations.

Institut National de la Recherche Agronomique, Paris

7. MILLER J.T. ed. (1971) *Instrument manual, 4th edition.*

Laboratory and workshop instruments mainly; principles and design. Several sections are relevant to ecology and micrometeorology, e.g. section 6 on air pollution, 7 on humidity, 8 on surface temperature.

United Trade Press, London.

Radio-telemetry

1. CACERES C.A. ed. (1965) *Bio-medical telemetry.*

Academic Press, New York.

2. MACKAY R.S. (1968) *Bio-medical telemetry: sensing and transmitting biological information for animals and man.*

Wiley, New York.

3. SLATER L. ed. (1963) *Bio-telemetry: the use of telemetry in animal behaviour and physiology in relation to ecological problems.*

Proc. Interdisciplinary Conference.

Macmillan, New York.

4. BARWICK R.E. & FULLAGAR P.J. (1967) *A bibliography of radio telemetry in biological studies.*

Proc. ecol. Soc. Aust. 2, 27-49.

Notes